# Data Science Solutions with Python

## Fast and Scalable Models Using Keras, PySpark MLlib, H2O, XGBoost, and Scikit-Learn

Tshepo Chris Nokeri

Apress®

*Data Science Solutions with Python: Fast and Scalable Models Using Keras, PySpark MLlib, H2O, XGBoost, and Scikit-Learn*

Tshepo Chris Nokeri
Pretoria, South Africa

ISBN-13 (pbk): 978-1-4842-7761-4
https://doi.org/10.1007/978-1-4842-7762-1

ISBN-13 (electronic): 978-1-4842-7762-1

Managing Director, Apress Media LLC: Welmoed Spahr
Acquisitions Editor: Celestin Suresh John
Development Editor: Laura Berendson
Coordinating Editor: Aditee Mirashi

Cover designed by eStudioCalamar

Cover image designed by Freepik (www.freepik.com)

Distributed to the book trade worldwide by Springer Science+Business Media New York, 1 New York Plaza, Suite 4600, New York, NY 10004-1562, USA. Phone 1-800-SPRINGER, fax (201) 348-4505, e-mail orders-ny@springer-sbm.com, or visit www.springeronline.com. Apress Media, LLC is a California LLC and the sole member (owner) is Springer Science + Business Media Finance Inc (SSBM Finance Inc). SSBM Finance Inc is a Delaware corporation.

For information on translations, please e-mail booktranslations@springernature.com; for reprint, paperback, or audio rights, please e-mail bookpermissions@springernature.com.

Apress titles may be purchased in bulk for academic, corporate, or promotional use. eBook versions and licenses are also available for most titles. For more information, reference our Print and eBook Bulk Sales web page at http://www.apress.com/bulk-sales.

Any source code or other supplementary material referenced by the author in this book is available to readers on GitHub via the book's product page, located at www.apress.com/9781484277614. For more detailed information, please visit http://www.apress.com/source-code.

Printed on acid-free paper

*I dedicate this book to my family and everyone who has merrily played influential roles in my life.*

# Table of Contents

# About the Author

**Tshepo Chris Nokeri** harnesses advanced analytics and artificial intelligence to foster innovation and optimize business performance. In his work, he delivered complex solutions to companies in the mining, petroleum, and manufacturing industries. He earned a Bachelor's degree in Information Management and then graduated with an honour's degree in Business Science from the University of the Witwatersrand, on a TATA Prestigious Scholarship and a Wits Postgraduate Merit Award. He was also unanimously awarded the Oxford University Press Prize. He is the author of *Data Science Revealed, Implementing Machine Learning in Finance*, and *Econometrics and Data Science*, all published by Apress.

# About the Technical Reviewer

**Joos Korstanje** is a data scientist with over five years of industry experience in developing machine learning tools, a large part of which has been forecasting models. He currently works at Disneyland Paris, where he develops machine learning for a variety of tools. His experience in writing and teaching have motivated him to contribute to this book on advanced forecasting with Python.

# Acknowledgments

Writing a single-authored book is demanding, but I received firm support and active encouragement from my family and dear friends. Many heartfelt thanks to the Apress Publishing team for all their support throughout the writing and editing processes. Lastly, my humble thanks to all of you for reading this; I earnestly hope you find it helpful.

# Introduction

This book covers the in-memory, distributed cluster computing framework called PySpark, the machine learning framework platforms called Scikit-Learn, PySpark MLlib, H2O, and XGBoost, and the deep learning framework known as Keras. After reading this book, you will be able to apply supervised and unsupervised learning to solve practical and real-world data problems. In this book, you will learn how to engineer features, optimize hyperparameters, train and test models, develop pipelines, and automate the machine learning process.

To begin, the book carefully presents supervised and unsupervised ML and DL models and examines big data frameworks and machine learning and deep learning frameworks. It also discusses the parametric model called Generalized Linear Model and a survival regression model known as the Cox Proportional Hazards model and Accelerated Failure Time (AFT). It presents a binary classification model called Logistic Regression and an ensemble model called Gradient Boost Trees. It also introduces DL and an artificial neural network, the Multilayer Perceptron (MLP) classifier. It describes a way of performing cluster analysis using the k-means model. It explores dimension reduction techniques like Principal Components Analysis and Linear Discriminant Analysis and concludes by unpacking automated machine learning.

The book targets intermediate data scientists and machine learning engineers who want to learn how to apply key big data frameworks, as well as ML and DL frameworks. Before exploring the contents of this book, be sure that you understand basic statistics, Python programming, probability theories, and predictive analytics.

The books uses Anaconda (an open source distribution of Python programming) for the examples. The following list highlights some of the Python libraries that this book covers.

- Pandas for data structures and tools.

- PySpark for in-memory, cluster computing.

- XGBoost for gradient boosting and survival regression analysis.

- Auto-Sklearn, Tree-based Pipeline Optimization Tool (TPOT), Hyperopt-Sklearn, and H2O for AutoML.

- Scikit-Learn for building and validating key machine learning algorithms.

- Keras for high-level frameworks for deep learning.

- H2O for driverless machine learning.

- Lifelines for survival analysis.

- NumPy for arrays and matrices.

- SciPy for integrals, solving differential equations, and optimization.

- Matplotlib and Seaborn for recognized plots and graphs.

# CHAPTER 1

# Exploring Machine Learning

This chapter introduces the best machine learning methods and specifies the main differences between supervised and unsupervised machine learning. It also discusses various applications of both.

Machine learning has been around for a long time; however, it has recently gained widespread recognition. This is because of the increased computational power of modern computer systems and the ease of access to open source platforms and frameworks. Machine learning involves inducing computer systems with intelligence by implementing various programming and statistical techniques. It draws from fields such as statistics, computational linguistics, and neuroscience, among others. It also applies modern statistics and basic programming. It enables developers to develop and deploy intelligent computer systems and create practical and reliable applications.

## Exploring Supervised Methods

Supervised learning involves feeding a machine learning algorithm with a large data set, through which it learns the desired output. The main machine learning methods are linear, nonlinear, and ensemble. The following section familiarizes you with the parametric method, also known as the *linear regression method*. See Table 1-1.

Linear regression methods expect data to follow a Gaussian distribution. The ordinary least-squares method is the standard linear method; it minimizes the error in terms and determines the slope and intercept to fit the data to a straight line (where the dependent feature is continuous).

Furthermore, the ordinary least-squares method determines the extent of the association between the features. It also assumes linearity, no autocorrelation in residuals, and no multicollinearity. In the real world, data hardly ever comes from a

1

© Tshepo Chris Nokeri 2022
T. C. Nokeri, *Data Science Solutions with Python*, https://doi.org/10.1007/978-1-4842-7762-1_1

normal distribution; the method struggles to explain the variability, resulting in an under-fitted or over-fitted model. To address this problem, we introduce a penalty term to the equation to alter the model's performance.

***Table 1-1.*** *Types of Parametric Methods*

| Model | Description |
|---|---|
| Linear regression method | Applied when there is one dependent feature (continuous feature) and an independent feature (continuous feature of categorical). The main linear regression methods are GLM, Ridge, Lasso, Elastic Net, etc. |
| Survival regression method | Applied to time-event-related censored data, where the dependent feature is categorical and the independent feature is continuous. |
| Time series analysis method | Applied to uncovering patterns in sequential data and forecasting future instances. Principal time series models include the ARIMA model, SARIMA, Additive model, etc. |

# Exploring Nonlinear Models

Nonlinear (classification) methods differentiate classes (also called categories or labels)—see Table 1-2. When there are two classes in the dependent feature, you'll implement binary classification methods. When there are more than two classes, you'll implement multiclass classification methods. There are innumerable functions for classification, including sigmoid, tangent hyperbolic, and kernel, among others. Their application depends on the context. Subsequent chapters will cover some of these functions.

***Table 1-2.*** *Varying Nonlinear Models*

| Model | Description |
| --- | --- |
| Binary classification method | Applied when the categorical feature has only two possible outcomes. The popular binary classifier is the logistic regression model. |
| Multiclass classification method | Applied when the categorical feature has more than two possible outcomes. The main multi-class classifier is the Linear Discriminant Analysis model (which can also be used for dimension reduction). |
| Survival classification | Applied when you're computing the probabilities of an event occurring using a categorical feature. |

# Exploring Ensemble Methods

Ensemble methods enable you to uncover linearity and nonlinearity. The main ensembler is the random forest trees, which is often computationally demanding. In addition, its performance depends on a variety of factors like the depth, iterations, data splitting, and boosting.

# Exploring Unsupervised Methods

In contrast to supervised machine learning, unsupervised learning contains no ground truth. You expose the model to all the sample data, allowing it to guess about the pattern of the data. The most common unsupervised methods are cluster-related methods.

# Exploring Cluster Methods

Cluster methods are convenient for assembling common values in the data; they enable you to uncover patterns in both structured and unstructured data. Table 1-3 highlights a few cluster methods that you can use to guess the pattern of the data.

***Table 1-3.*** *Varying Cluster Models*

| Techniques | Description |
|---|---|
| Centroid clustering | Applied to determine the center of the data and draw data points toward the center. The main centroid clustering method is the k-means method. |
| Density clustering | Applied to determine where the data is concentrated. The main density clustering model is the DBSCAN method. |
| Distribution clustering | Identifies the probability of data points belonging to a cluster based on some distribution. The main distribution clustering method is the Gaussian Mixture method. |

# Exploring Dimension Reduction

Dimension reducers help determine the extent to which factors or components elucidate related changes in data. Table 1-4 provides an overview of the chief dimension reducers.

***Table 1-4.*** *Main Dimension Reducers*

| Technique | Description |
|---|---|
| Factor analysis | Applied to determine the extent to which factors elucidate related changes of features in the data. |
| Principal component analysis | Applied to determine the extent to which factors elucidate related changes of features in the data. |

# Exploring Deep Learning

Deep learning extends machine learning in that it uses artificial neural networks to solve complex problems. You can use deep learning to discover patterns in big data. Deep learning uses artificial neural networks to discover patterns in complex data. The networks use an approach similar to that of animals, whereby neurons (nodes) that receive input data from the environment scan the input and pass it to neurons in successive layers to arrive at some output (which is understanding the complexity in the data). There are several artificial neural networks that you can use to determine patterns of behavior of a phenomenon, depending on the context. Table 1-5 highlights the chief neural networks.

***Table 1-5.***  *Varying Neural Networks*

| Network | Description |
| --- | --- |
| Restricted Boltzmann Machine (RBM) | The most common neural network that contains only the hidden and visible layers. |
| Multilayer Perceptron (MLP) | A neural network that prolongs a restricted Boltzmann machine with input, hidden and output layers. |
| Recurrent Neural Network (RNN) | Serves as a sequential modeler. |
| Convolutional Neural Network (CNN) | Serves as a dimension reducer and classifier. |

# Conclusion

This chapter covered two ways in which machines learn—via supervised and unsupervised learning. It began by explaining supervised machine learning and discussing the three types of supervised learning methods and their applications. It then covered unsupervised learning techniques, dimension reduction, and cluster analysis.

**CHAPTER 2**

# Big Data, Machine Learning, and Deep Learning Frameworks

This chapter carefully presents the big data framework used for parallel data processing called Apache Spark. It also covers several machine learning (ML) and deep learning (DL) frameworks useful for building scalable applications. After reading this chapter, you will understand how big data is collected, manipulated, and examined using resilient and fault-tolerant technologies. It discusses the Scikit-Learn, Spark MLlib, and XGBoost frameworks. It also covers a deep learning framework called Keras. It concludes by discussing effective ways of setting up and managing these frameworks.

Big data frameworks support parallel data processing. They enable you to contain big data across many clusters. The most popular big data framework is Apache Spark, which is built on the Hadoop framework.

## Big Data

Big data means different things to different people. In this book, we define big data as large amounts of data that we cannot adequately handle and manipulate using classic methods. We must undoubtedly use scalable frameworks and modern technologies to process and draw insight from this data. We typically consider data "big" when it cannot fit within the current in-memory storage space. For instance, if you have a personal computer and the data at your disposal exceeds your computer's storage capacity, it's big data. This equally applies to large corporations with large clusters of storage space. We often speak about big data when we use a stack with Hadoop/Spark.

© Tshepo Chris Nokeri 2022
T. C. Nokeri, *Data Science Solutions with Python*, https://doi.org/10.1007/978-1-4842-7762-1_2

# Big Data Features

The features of big data are described as the four Vs—velocity, volume, variety, and veracity. Table 2-1 highlights these features of big data.

***Table 2-1.*** *Big Data Features*

| Element | Description |
| --- | --- |
| Velocity | Modern technologies and improved connectivity enable you to generate data at an unprecedented speed. Characteristics of velocity include batch data, near or real-time data, and streams. |
| Volume | The scale at which data increases. The nature of data sources and infrastructure influence the volume of data. Characteristics of the volume include exabyte, zettabyte, etc. |
| Variety | Data can come from unique sources. Modern technological devices leave digital footprints here and there, which increase the number of sources from which businesses and people can get data. Characteristics of variety include the structure and complexity of the data. |
| Veracity | Data must come from reliable sources. Also, it must be of high quality, consistent, and complete. |

# Impact of Big Data on Business and People

Without a doubt, big data affects the way we think and do business. Data-driven organizations typically establish the basis for evidence-based management. Big data involves measuring the key aspects of the business using quantitative methods. It helps support decision-making. The next sections discuss ways in which big data affects businesses and people.

# Better Customer Relationships

Insights from big data help manage customer relationships. Organizations with big data about their customers can study customers' behavioral patterns and use descriptive analytics to drive customer-management strategies.

# Refined Product Development

Data-driven organizations use big data analytics and predictive analytics to drive product development. and management strategies. This approach is useful for incremental and iterative delivery of applications.

# Improved Decision-Making

When a business has big data, it can use it to uncover complex patterns of a phenomenon to influence strategy. This approach helps management make well-informed decisions based on evidence, rather than on subjective reasoning. Data-driven organizations foster a culture of evidence-based management.

We also use big data in fields like life sciences, physics, economics, and medicine. There are many ways in which big data affects the world. This chapter does not consider all factors. The next sections explain big data warehousing and ETL activities.

# Big Data Warehousing

Over the past few decades, organizations have invested in on-premise databases, including Microsoft Access, Microsoft SQL Server, SAP Hana, Oracle Database, and many more. There has recently been widespread adoption of cloud databases like Microsoft Azure SQL and Oracle XE. There are also standard big data (distributed) databases like Cassandra and HBase, among others. Businesses are shifting toward scalable cloud-based databases to harness benefits associated with increasing computational power, fault-tolerant technologies, and scalable solutions.

# Big Data ETL

Although there have been significant advances in database management, the way that people manipulate data from databases remains the same. Extracting, transforming, and loading (ETL) still play an integral part in analysis and reporting. Table 2-2 discusses ETL activities.

***Table 2-2.*** *ETL Activities*

| Activity | Description |
| --- | --- |
| Extract | Involves getting data from some database. |
| Transforming | Involves converting data from a database into a suitable format for analysis and reporting |
| Loading | Involves warehousing data in a database management system. |

To perform ETL activities, you must use a query language. The most popular query language is SQL (Standard Query Language). There are other query languages that emerged with the open source movement, such as HiveQL and BigQuery. The Python programming language supports SQL. Python frameworks can connect to databases by implementing libraries, such as SQLAlchemy, pyodbc, SQLite, SparkSQL, and pandas, among others.

# Big Data Frameworks

Big data frameworks enable developers to collect, manage, and manipulate distributed data. Most open source big data frameworks use in-memory cluster computing. The most popular frameworks include Hadoop, Spark, Flink, Storm, and Samza. This book uses PySpark to perform ETL activities, explore data, and build machine learning pipelines.

## Apache Spark

Apache Spark executes in-memory cluster computing. It enables developers to build scalable applications using Java, Scala, Python, R, and SQL. It includes cluster components like the driver, cluster manager, and executor. You can use it as a standalone cluster manager or on top of Mesos, Hadoop, YARN, or Baronets. You can use it to access data in the Hadoop File System (HDFS), Cassandra, HBase, and Hive, among other data sources. The Spark data structure is considered a resilient distributed data set. This book introduces a framework that integrates both Python and Apache Spark (PySpark). The book uses it to operate Spark MLlib. To understand this framework, you first need to grasp the idea behind resilient distributed data sets.

# Resilient Distributed Data Sets

Resilient Distributed Data Sets (RDDs) are immutable elements for parallelizing data or for transforming existing data. Chief RDD operations include transformation and actions. We store them in any storage supported by Hadoop. For instance, in a Hadoop Distributed File System (HDF), Cassandra, HBase, Amazon S3, etc.

# Spark Configuration

Areas of Spark configuration include Spark properties, environment variables, and logging. The default configuration directory is SPARK_HOME/conf.

You can install the findspark library in your environment using pip install findspark and install the pyspark library using pip install pyspark.

Listing 2-1 prepares the PySpark framework using the findspark framework.

***Listing 2-1.*** Prepare the PySpark Framework

```
import findspark as initiate_pyspark
initiate_pyspark.init("filepath\spark-3.0.0-bin-hadoop2.7")
```

Listing 2-2 stipulates the PySpark app using the SparkConf() method.

***Listing 2-2.*** Stipulate the PySpark App

```
from pyspark import SparkConf
pyspark_configuration = SparkConf().setAppName("pyspark_linear_method").
setMaster("local")
```

Listing 2-3 prepares the PySpark session with the SparkSession() method.

***Listing 2-3.*** Prepare the Spark Session

```
from pyspark.sql import SparkSession
pyspark_session = SparkSession(pyspark_context)
```

# Spark Frameworks

Spark frameworks extend the core of the Spark API. There are four main Spark frameworks—SparkSQL, Spark Streaming, Spark MLlib, and GraphX.

## SparkSQL

SparkSQL enables you to use relational query languages like SQL, HiveQL, and Scala. It includes a schemaRDD that has row objects and schema. You create it using an existing RDD, parquet file, or JSON data set. You execute the Spark Context to create a SQL context.

## Spark Streaming

Spark streaming is a scalable streaming framework that supports Apache Kafka, Apache Flume, HDFS, and Apache Kensis, etc. It processes input data using DStream in small batches you push using HDFS, databases, and dashboards. Recent versions of Python do not support Spark Streaming. Consequently, we do not cover the framework in this book. You can use a Spark Streaming application to read input from any data source and store a copy of the data in HDFS. This allows you to build and launch a Spark Streaming application that processes incoming data and runs an algorithm on it.

## Spark MLlib

MLlib is an ML framework that allows you to develop and test ML and DL models. In Python, the frameworks work hand-in-hand with the NumPy framework. Spark MLlib can be used with several Hadoop data sources and incorporated alongside Hadoop workflows. Common algorithms include regression, classification, clustering, collaborative filtering, and dimension reduction. Key workflow utilities include feature transformation, standardization and normalization, pipeline development, model evaluation, and hyperparameter optimization.

## GraphX

GraphX is a scalable and fault-tolerant framework for iterative and fast graph parallel computing, social networks, and language modeling. It includes graph algorithms such as PageRank for estimating the importance of each vertex in a graph, Connected Components for labeling connected components of the graph with the ID of its lowest-numbered vertex, and Triangle Counting for finding the number of triangles that pass through each vertex.

# ML Frameworks

To solve ML problems, you need to have a framework that supports building and scaling ML models. There is no shortage of ML models – there are innumerable frameworks for ML. There are several ML frameworks that you can use. Subsequent chapters cover frameworks like Scikit-Learn, Spark MLlib, H2O, and XGBoost.

## Scikit-Learn

The Scikit-Learn framework includes ML algorithms like regression, classification, and clustering, among others. You can use it with other frameworks such as NumPy and SciPy. It can perform most of the tasks required for ML projects like data processing, transformation, data splitting, normalization, hyperparameter optimization, model development, and evaluation. Scikit-Learn comes with most distribution packages that support Python. Use `pip install sklearn` to install it in your Python environment.

## H2O

H2O is an ML framework that uses a driverless technology. It enables you to accelerate the adoption of AI solutions. It is very easy to use, and it does not require any technical expertise. Not only that, but it supports numerical and categorical data, including text. Before you train the ML model, you must first load the data into the H2O cluster. It supports CSV, Excel, and Parquet files. Default data sources include local file systems, remote files, Amazon S3, HDFS, etc. It has ML algorithms like regression, classification, cluster analysis, and dimension reduction. It can also perform most tasks required for ML projects like data processing, transformation, data splitting, normalization, hyperparameter optimization, model development, checking pointing, evaluation, and productionizing. Use `pip install h2o` to install the package in your environment.

Listing 2-4 prepares the H2O framework.

***Listing 2-4.*** Initializing the H2O Framework

```
import h2o
h2o.init()
```

# XGBoost

XGBoost is an ML framework that supports programming languages, including Python. It executes gradient-boosted models that are scalable, and learns fast parallel and distributed computing without sacrificing memory efficiency. Not only that, but it is an ensemble learner. As mentioned in Chapter a, ensemble learners can solve both regression and classification problems. XGBoost uses boosting to learn from the errors committed in the preceding trees. It is useful when tree-based models are overfitted. Use `pip install xgboost` to install the model in your Python environment.

# DL Frameworks

DL frameworks provide a structure that supports scaling artificial neural networks. You can use it stand-alone or with other models. It typically includes programs and code frameworks. Primary DL frameworks include TensorFlow, PyTorch, Deeplearning4j, Microsoft Cognitive Toolkit (CNTK), and Keras.

## Keras

Keras is a high-level DL framework written using Python; it runs on top of an ML platform known as TensorFlow. It is effective for rapid prototyping of DL models. You can run Keras on Tensor Processing Units or on massive Graphic Processing Units. The main Keras APIs include models, layers, and callbacks. Chapter 7 covers this framework. Execute `pip install Keras` and `pip install tensorflow` to use the Keras framework.

# CHAPTER 3

# Linear Modeling with Scikit-Learn, PySpark, and H2O

This introductory chapter explains the ordinary least-squares method and executes it with the main Python frameworks (i.e., Scikit-Learn, Spark MLlib, and H2O). It begins by explaining the underlying concept behind the method.

## Exploring the Ordinary Least-Squares Method

The ordinary least-squares method is used with data that has an output feature that is not confined (it's continuous). This method expects normality and linearity, and there must be an absence of autocorrelation in the error of terms (also called *residuals*) and multicollinearity. It is also highly prone to abnormalities in the data, so you might want to use alternative methods like Ridge, Lasso, and Elastic Net if this method does not serve you well.

Listing 3-1 attains the necessary data from a Microsoft CSV file.

***Listing 3-1.*** Attain the Data

```
import pandas as pd
df = pd.read_csv(r"filepath\WA_Fn-UseC_-Marketing_Customer_Value_
Analysis.csv")
```

Listing 3-2 stipulates the names of columns to drop and then executes the `drop()` method. It then stipulates axes as `columns` in order to drop the unnecessary columns in the data.

15

**Listing 3-2.** Drop Unnecessary Features in the Data

```
drop_column_names = df.columns[[0, 6]]
initial_data = df.drop(drop_column_names, axis="columns")
```

Listing 3-3 attains the dummy values for the categorical features in this data.

**Listing 3-3.** Attain Dummy Features

```
initial_data.iloc[::, 0] = pd.get_dummies(initial_data.iloc[::, 0])
initial_data.iloc[::, 2] = pd.get_dummies(initial_data.iloc[::, 2])
initial_data.iloc[::, 3] = pd.get_dummies(initial_data.iloc[::, 3])
initial_data.iloc[::, 4] = pd.get_dummies(initial_data.iloc[::, 4])
initial_data.iloc[::, 5] = pd.get_dummies(initial_data.iloc[::, 5])
initial_data.iloc[::, 6] = pd.get_dummies(initial_data.iloc[::, 6])
initial_data.iloc[::, 7] = pd.get_dummies(initial_data.iloc[::, 7])
initial_data.iloc[::, 8] = pd.get_dummies(initial_data.iloc[::, 8])
initial_data.iloc[::, 9] = pd.get_dummies(initial_data.iloc[::, 9])
initial_data.iloc[::, 15] = pd.get_dummies(initial_data.iloc[::, 15])
initial_data.iloc[::, 16] = pd.get_dummies(initial_data.iloc[::, 16])
initial_data.iloc[::, 17] = pd.get_dummies(initial_data.iloc[::, 17])
initial_data.iloc[::, 18] = pd.get_dummies(initial_data.iloc[::, 18])
initial_data.iloc[::, 20] = pd.get_dummies(initial_data.iloc[::, 20])
initial_data.iloc[::, 21] = pd.get_dummies(initial_data.iloc[::, 21])
```

Listing 3-4 outlines the independent and dependent features.

**Listing 3-4.** Outline the Features

```
import numpy as np
int_x = initial_data.iloc[::,0:19]
fin_x = initial_data.iloc[::,19:21]
x_combined = pd.concat([int_x, fin_x], axis=1)
x = np.array(x_combined)
y = np.array(initial_data.iloc[::,19])
```

# Scikit-Learn in Action

Listing 3-5 randomly divides the dataframe.

***Listing 3-5.*** Randomly Divide the Dataframe

```
from sklearn.model_selection import train_test_split
x_train, x_test, y_train, y_test = train_test_split(x, y, test_size=0.2,
random_state=0)
```

Listing 3-6 scales the independent features.

***Listing 3-6.*** Scale the Independent Features

```
from sklearn.preprocessing import StandardScaler
sk_standard_scaler = StandardScaler()
sk_standard_scaled_x_train = sk_standard_scaler.fit_transform(x_train)
sk_standard_scaled_x_test = sk_standard_scaler.transform(x_test)
```

Listing 3-7 executes the Scikit-Learn ordinary least-squares regression method.

***Listing 3-7.*** Execute the Scikit-Learn Ordinary Least-Squares Regression Method

```
from sklearn.linear_model import LinearRegression
sk_linear_model = LinearRegression()
sk_linear_model.fit(sk_standard_scaled_x_train, y_train)
```

Listing 3-8 determines the best hyperparameters for the Scikit-Learn ordinary least-squares regression method.

***Listing 3-8.*** Determine the Best Hyperparameters for the Scikit-Learn Ordinary Least-Squares Regression Method

```
from sklearn.preprocessing import GridSearchCV
sk_linear_model_param = {'fit_intercept':[True,False]}
sk_linear_model_param_mod  = GridSearchCV(estimator=sk_linear_model,
param_grid=sk_linear_model_param, n_jobs=-1)
sk_linear_model_param_mod.fit(sk_standard_scaled_x_train, y_train)
```

```
print("Best OLS score: ", sk_linear_model_param_mod.best_score_)
print("Best OLS parameter: ", sk_linear_model_param_mod.best_params_)

Best OLS score:  1.0
Best OLS parameter:  {'fit_intercept': True}
```

Listing 3-9 executes the Scikit-Learn ordinary least-squares regression method.

***Listing 3-9.*** Execute the Scikit-Learn Ordinary Least-Squares Regression Method

```
sk_linear_model = LinearRegression(fit_intercept= True)
sk_linear_model.fit(sk_standard_scaled_x_train, y_train)
sk_linear_model_param_mod.best_params_)
```

Listing 3-10 computes the Scikit-Learn ordinary least-squares regression method's intercept.

***Listing 3-10.*** Compute the Scikit-Learn Ordinary Least-Squares Regression Method's Intercept

```
print(sk_linear_model.intercept_)
433.0646521131769
```

Listing 3-11 computes the Scikit-Learn ordinary least-squares regression method's coefficients.

***Listing 3-11.*** Compute the Scikit-Learn Ordinary Least-Squares Regression Method's Coefficients

```
print(sk_linear_model.coef_)
[-6.15076155e-15  2.49798076e-13 -1.95573220e-14 -1.90089677e-14
 -5.87187344e-14  2.50923806e-14 -1.05879478e-13  1.53591400e-14
 -1.82507711e-13 -7.86327034e-14  4.17629484e-13  1.28923537e-14
  6.52911311e-14 -5.28069778e-14 -1.57900159e-14 -6.74040176e-14
 -9.28427833e-14  5.03132848e-14 -8.75978166e-15  2.90235705e+02
 -9.55950515e-14]
```

Listing 3-12 computes the ordinary least-squares regression method's predictions.

***Listing 3-12.*** Compute the Scikit-Learn Ordinary Least-Squares Regression Method's Predictions

```
sk_yhat = sk_linear_model.predict(sk_standard_scaled_x_test)
```

Listing 3-13 assesses the Scikit-Learn ordinary least-squares method (see Table 3-1).

***Listing 3-13.*** Assess the Scikit-Learn Ordinary Least-Squares Method

```
from sklearn import metrics
sk_mean_ab_error = metrics.mean_absolute_error(y_test, sk_yhat)
sk_mean_sq_error = metrics.mean_squared_error(y_test, sk_yhat)
sk_root_sq_error = np.sqrt(sk_mean_sq_error)
sk_determinant_coef = metrics.r2_score(y_test, sk_yhat)
sk_exp_variance = metrics.explained_variance_score(y_test, sk_yhat)
sk_linear_model_ev = [[sk_mean_ab_error, sk_mean_sq_error, sk_root_sq_
                    error, sk_determinant_coef, sk_exp_variance]]
sk_linear_model_assessment = pd.DataFrame(sk_linear_model_ev, index =
["Estimates"], columns = ["Sk mean absolute error", "Sk mean squared error",
                    "Sk root mean squared error", "Sk determinant
                    coefficient", "Sk variance score"])
sk_linear_model_assessment
y = np.array(initial_data.iloc[::,19])
```

***Table 3-1.*** *Assessment of the Scikit-Learn Ordinary Least-Squares Method*

|  | Sk Mean Absolute Error | Sk Mean Squared Error | Sk Root Mean Squared Error | Sk Determinant Coefficient | Sk Variance Score |
|---|---|---|---|---|---|
| **Estimates** | 9.091189e-13 | 1.512570e-24 | 1.229866e-12 | 1.0 | 1.0 |

Table 3-1 shows that the Scikit-Learn ordinary least-squares method explains the entire variability.

# PySpark in Action

This section executes and assesses the ordinary least-squares method with the PySpark framework. Listing 3-14 prepares the PySpark framework with the `findspark` framework.

***Listing 3-14.*** Prepare the PySpark Framework

```
import findspark as initiate_pyspark
initiate_pyspark.init("filepath\spark-3.0.0-bin-hadoop2.7")
```

Listing 3-15 stipulates the PySpark app with the `SparkConf()` method.

***Listing 3-15.*** Stipulate the PySpark App

```
from pyspark import SparkConf
pyspark_configuration = SparkConf().setAppName("pyspark_linear_method").
setMaster("local")
```

Listing 3-16 prepares the PySpark session with the `SparkSession()` method.

***Listing 3-16.*** Prepare the Spark Session

```
from pyspark.sql import SparkSession
pyspark_session = SparkSession(pyspark_context)
```

Listing 3-17 changes the `pandas` dataframe created earlier in this chapter to a PySpark dataframe using the `createDataFrame()` method.

***Listing 3-17.*** Change Pandas Dataframe to a PySpark Dataframe

```
pyspark_initial_data = pyspark_session.createDataFrame(initial_data)
```

Listing 3-18 creates a list for independent features and a string for the dependent feature. It converts data using the `VectorAssembler()` method for modeling with the PySpark framework.

***Listing 3-18.*** Transform the Data

```
x_list = list(x_combined.columns)
y_list = initial_data.columns[19]
from pyspark.ml.feature import VectorAssembler
```

```
pyspark_data_columns = x_list
pyspark_vector_assembler = VectorAssembler(inputCols=pyspark_data_columns,
outputCol="variables")
pyspark_data = pyspark_vector_assembler.transform(pyspark_initial_data)
```

Listing 3-19 divides the data using the randomSplit() method.

***Listing 3-19.*** Divide the Dataframe

```
(pyspark_training_data, pyspark_test_data) = pyspark_data.
randomSplit([.8,.2])
```

Listing 3-20 executes the PySpark ordinary least-squares regression method.

***Listing 3-20.*** Execute the PySpark Ordinary Least-Squares Regression Method

```
from pyspark.ml.regression import LinearRegression
pyspark_linear_model = LinearRegression(labelCol=y_list,
featuresCol=pyspark_data.columns[-1])
pyspark_fitted_linear_model = pyspark_linear_model.fit(pyspark_training_data)
```

Listing 3-21 computes the PySpark ordinary least-squares regression method's predictions.

***Listing 3-21.*** Compute the PySpark Ordinary Least-Squares Regression Method's Predictions

```
pyspark_yhat = pyspark_fitted_linear_model.transform(pyspark_test_data)
```

Listing 3-22 assesses the PySpark ordinary least-squares method.

***Listing 3-22.*** Assess the PySpark Ordinary Least-Squares Method

```
pyspark_linear_model_assessment = pyspark_fitted_linear_model.summary
print("PySpark root mean squared error", pyspark_linear_model_assessment.
rootMeanSquaredError)
print("PySpark determinant coefficient", pyspark_linear_model_assessment.r2)

PySpark root mean squared error 2.0762306526480097e-13
PySpark determinant coefficient 1.0
```

# H2O in Action

This section executes and assesses the ordinary least-squares method with the H2O framework.

Listing 3-23 prepares the H2O framework.

***Listing 3-23.*** Prepare the H2O Framework

```
import h2o as initialize_h2o
initialize_h2o.init()
```

Listing 3-24 changes the pandas dataframe to an H2O dataframe.

***Listing 3-24.*** Change the Pandas Dataframe to an H2O Dataframe

```
h2o_data = initialize_h2o.H2OFrame(initial_data)
```

Listing 3-25 outlines the independent and dependent features.

***Listing 3-25.*** Outline Features

```
y = y_list
x = h2o_data.col_names
x.remove(y_list)
```

Listing 3-26 randomly divides the data.

***Listing 3-26.*** Randomly Divide the Dataframe

```
h2o_training_data, h2o_validation_data, h2o_test_data = h2o_data.split_
frame(ratios=[.8,.1])
```

Listing 3-27 executes the H2O ordinary least-squares regression method.

***Listing 3-27.*** Execute the H2O Ordinary Least-Squares Regression Method

```
from h2o.estimators import H2OGeneralizedLinearEstimator
h2o_linear_model = H2OGeneralizedLinearEstimator(family="gaussian")
h2o_linear_model.train(x=x,y=y,training_frame=h2o_training_data,validation_
frame=h2o_validation_data)
```

Listing 3-28 computes the H2O ordinary least-squares method's predictions.

***Listing 3-28.*** H2O Ordinary Least-Squares Method Executed Predictions

```
h2o_yhat = h2o_linear_model.predict(h2o_test_data)
```

Listing 3-29 computes the H2O ordinary least-squares method's standardized coefficients (see Figure 3-1).

***Listing 3-29.*** H2O Ordinary Least-Squares Method's Standardized Coefficients

```
h2o_linear_model_std_coefficients = h2o_linear_model.std_coef_plot()
h2o_linear_model_std_coefficients
```

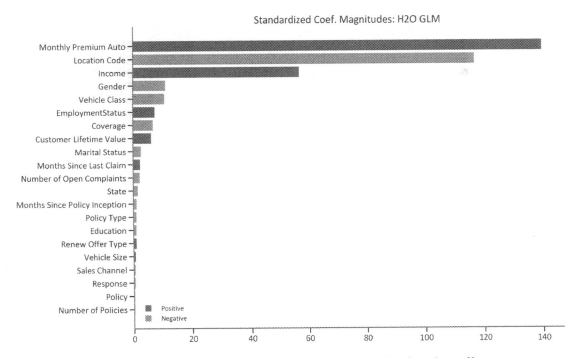

***Figure 3-1.*** *H2O ordinary least-squares method's standardized coefficients*

Listing 3-30 computes the H2O ordinary least-squares method's partial dependency (see Figure 3-2).

**Listing 3-30.** H2O Ordinary Least-Squares Method's Partial Dependency

```
h2o_linear_model_dependency_plot = h2o_linear_model.partial_plot
(data = h2o_data, cols = list(initial_data.columns[[0,19]]), server=False,
plot = True)
h2o_linear_model_dependency_plot
```

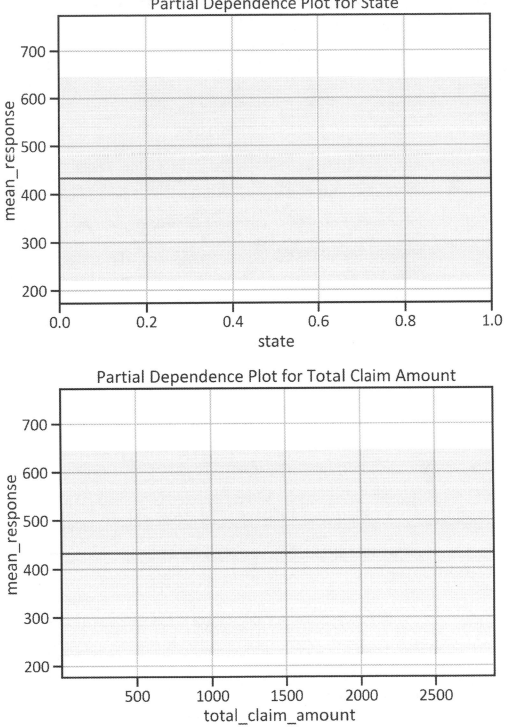

**Figure 3-2.** *H2O ordinary least-squares method's partial dependency*

Listing 3-31 arranges the features that are the most important to the H2O ordinary least-squares method in ascending order (see Figure 3-3).

**Listing 3-31.** H2O Ordinary Least-Squares Method's Feature Importance

```
h2o_linear_model_feature_importance = h2o_linear_model.varimp_plot()
h2o_linear_model_feature_importance
```

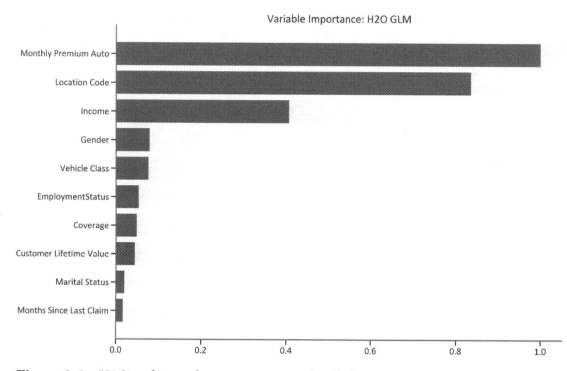

**Figure 3-3.** *H2O ordinary least-squares method's feature importance*

Listing 3-32 assesses the H2O ordinary least-squares method.

**Listing 3-32.** Assess H2O Ordinary Least-Squares Method

```
h2o_linear_model_assessment = h2o_linear_model.model_performance()
print(h2o_linear_model_assessment)
```

ModelMetricsRegressionGLM: glm
** Reported on train data. **

MSE: 24844.712331260016

```
RMSE: 157.6220553452467
MAE: 101.79904883889066
RMSLE: NaN
R^2: 0.7004468136072375
Mean Residual Deviance: 24844.712331260016
Null degrees of freedom: 7325
Residual degrees of freedom: 7304
Null deviance: 607612840.7465751
Residual deviance: 182012362.53881088
AIC: 94978.33944003603
```

Listing 3-33 improves the performance of the H2O ordinary least-squares method by specifying remove_collinear_columns as True.

**Listing 3-33.** Improve the Performance of the Ordinary Least-Squares Method

```
h2o_linear_model_collinear_removed = H2OGeneralizedLinearEstimator(family="
gaussian", lambda_ = 0,remove_collinear_columns = True)
h2o_linear_model_collinear_removed.train(x=x,y=y,training_frame=h2o_
training_data,validation_frame=h2o_validation_data)
```

Listing 3-34 assesses the H2O ordinary least-squares method.

**Listing 3-34.** Assess the H2O Ordinary Least-Squares Method

```
h2o_linear_model_collinear_removed_assessment = h2o_linear_model_collinear_
removed.model_performance()
print(h2o_linear_model_collinear_removed)
```

```
MSE: 23380.71864337616
RMSE: 152.9075493341521
MAE: 102.53007935777588
RMSLE: NaN
R^2: 0.7180982143647627
Mean Residual Deviance: 23380.71864337616
Null degrees of freedom: 7325
Residual degrees of freedom: 7304
Null deviance: 607612840.7465751
```

```
Residual deviance: 171287144.78137374
AIC: 94533.40762597627

ModelMetricsRegressionGLM: glm
** Reported on validation data. **

MSE: 25795.936313899092
RMSE: 160.6111338416459
MAE: 103.18677222520363
RMSLE: NaN
R^2: 0.7310558588001701
Mean Residual Deviance: 25795.936313899092
Null degrees of freedom: 875
Residual degrees of freedom: 854
Null deviance: 84181020.04623385
Residual deviance: 22597240.210975606
AIC: 11430.364002305443
```

# Conclusion

This chapter executed three key machine learning frameworks (Scikit-Learn, PySpark, and H2O) to model data and spawn a continuous output feature using a linear method. It also explored ways of assessing that method.

# CHAPTER 4

# Survival Analysis with PySpark and Lifelines

This chapter describes and executes several survival analysis methods using the main Python frameworks (i.e., Lifelines and PySpark). It begins by explaining the underlying concept behind the Cox Proportional Hazards model. It then introduces the accelerated failure time method.

## Exploring Survival Analysis

Survival methods are common in the manufacturing, insurance, and medical science fields. They are convenient for properly assessing risk when an independent investigation is carried out over long periods and subjects enter and leave at given times.

These methods are employed in this chapter to determine the probabilities of a machine failing and patients surviving an illness, among other applications. They are equally suitable for missing values (censored data).

## Exploring Cox Proportional Hazards Method

The Cox Proportional Hazards method is the best survival method for handling censored data with subjects that have related changes. It is similar to the Mantel-Haenszel method and expects the hazard rate to be a function of time. It states that the rate is a function of covariates. Equation 4-1 defines the Cox Proportional Hazards method.

$$\left(t, x_1, x_2, \ldots, x_p\right) = \left(t\right) exp\left(b_1 X_1 + b_2 X_2 + \ldots b_p X_p\right) \qquad \text{(Equation 4-1)}$$

Listing 4-1 attains the necessary data from a Microsoft Excel file.

© Tshepo Chris Nokeri 2022
T. C. Nokeri, *Data Science Solutions with Python*, https://doi.org/10.1007/978-1-4842-7762-1_4

***Listing 4-1.*** Attain the Data

```
import pandas as pd
initial_data = pd.read_excel(r"filepath\survival_data.xlsx", index_col=[0])
```

Listing 4-2 finds the ratio for dividing the data.

***Listing 4-2.*** Find the Ratio for Dividing the Data

```
int(initial_data.shape[0]) * 0.8
```

345.6

Listing 4-3 divides the data.

***Listing 4-3.*** Divide the Data

```
lifeline_training_data = initial_data.loc[:346]
lifeline_test_data = initial_data.loc[346:]
```

# Lifeline in Action

This section executes and assesses the Cox Proportional Hazards method with the Lifeline framework. Listing 4-4 executes the Lifeline Cox Proportional Hazards method.

***Listing 4-4.*** Execute the Lifeline Cox Proportional Hazards Method

```
from lifelines import CoxPHFitter
lifeline_cox_method = CoxPHFitter()
lifeline_cox_method.fit(lifeline_training_data, initial_data.columns[0],
initial_data.columns[1])
```

Listing 4-5 computes the test statistics (see Table 4-1) and assesses the Lifeline Cox Proportional Hazards method with a scaled Schoenfeld, which helps disclose any abnormalities in the residuals (see Figure 4-1).

***Listing 4-5.*** Compute the Lifeline Cox Proportional Hazards Method's Test Statistics and Residuals

```
lifeline_cox_method_test_statistics_schoenfeld = lifeline_cox_method.
check_assumptions(lifeline_training_data, show_plots=True)
lifeline_cox_method_test_statistics_schoenfeld
```

***Table 4-1.*** *Test Statistics for the Lifeline Cox Proportional Hazards Method*

|  |  | Test Statistic | p |
| --- | --- | --- | --- |
| Age | km | 10.53 | <0.005 |
|  | rank | 10.78 | <0.005 |
| Fin | km | 0.12 | 0.73 |
|  | rank | 0.14 | 0.71 |
| Mar | km | 0.18 | 0.67 |
|  | rank | 0.20 | 0.66 |
| Paro | km | 0.13 | 0.72 |
|  | rank | 0.11 | 0.74 |
| Prio | km | 0.49 | 0.48 |
|  | rank | 0.47 | 0.49 |
| Race | km | 0.34 | 0.56 |
|  | rank | 0.37 | 0.54 |
| Wexp | km | 11.91 | <0.005 |
|  | rank | 11.61 | <0.005 |

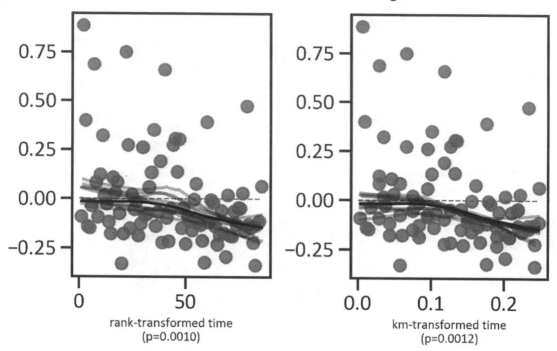

*Figure 4-1. Scaled Schoenfeld residuals of age*

Listing 4-6 determines the Lifeline Cox Proportional Hazards method's assessment summary (see Table 4-2).

***Listing 4-6.*** Compute the Assessment Summary

```
lifeline_cox_method_assessment_summary = lifeline_cox_method.
print_summary()
lifeline_cox_method_assessment_summary
```

***Table 4-2.*** *Summary of the Cox Proportional Hazards*

| Coef | Exp(coef) | Se(coef) | Coef Lower 95% | Coef Upper 95% | Exp(coef) Lower 95% | Exp(coef) Upper 95% | Z | P | -log2(p) |
|------|-----------|----------|------|------|------|------|------|------|------|
| Fin | -0.71 | 0.49 | 0.23 | -1.16 | -0.27 | 0.31 | 0.77 | -3.13 | <0.005 | 9.14 |
| Age | -0.03 | 0.97 | 0.02 | -0.08 | 0.01 | 0.93 | 1.01 | -1.38 | 0.17 | 2.57 |
| Race | 0.39 | 1.48 | 0.37 | -0.34 | 1.13 | 0.71 | 3.09 | 1.05 | 0.30 | 1.76 |
| Wexp | -0.11 | 0.90 | 0.24 | -0.59 | 0.37 | 0.56 | 1.44 | -0.45 | 0.65 | 0.62 |
| Mar | -1.15 | 0.32 | 0.61 | -2.34 | 0.04 | 0.10 | 1.04 | -1.90 | 0.06 | 4.11 |
| Paro | 0.07 | 1.07 | 0.23 | -0.37 | 0.51 | 0.69 | 1.67 | 0.31 | 0.76 | 0.40 |
| Prio | 0.10 | 1.11 | 0.03 | 0.04 | 0.16 | 1.04 | 1.17 | 3.24 | <0.005 | 9.73 |

Listing 4-7 determines the log test confidence interval for each feature in the data (see Figure 4-2).

***Listing 4-7.*** Execute the Lifeline Cox Proportional Hazards Method

```
lifeline_cox_log_test_ci = lifeline_cox_method.plot()
lifeline_cox_log_test_ci
```

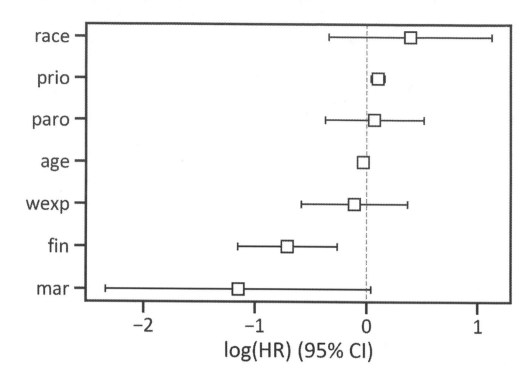

***Figure 4-2.*** *Log test confidence interval*

# Exploring the Accelerated Failure Time Method

The accelerated failure time method models the censored data with a log-linear function to describe the log of the survival time. Likewise, it also assumes each instance is independent.

# PySpark in Action

This section executes the accelerated failure time method with the PySpark framework.

Listing 4-8 runs the PySpark framework with the findspark framework.

***Listing 4-8.*** Prepare the PySpark Framework

```
import findspark as initiate_pyspark
initiate_pyspark.init("filepath\spark-3.0.0-bin-hadoop2.7")
```

Listing 4-9 stipulates the PySpark app using the SparkConf() method.

***Listing 4-9.*** Stipulate the PySpark App

```
from pyspark import SparkConf
pyspark_configuration = SparkConf().setAppName("pyspark_aft_method").
setMaster("local")
```

Listing 4-10 prepares the PySpark session using the SparkSession() method.

***Listing 4-10.*** Prepare the Spark Session

```
from pyspark.sql import SparkSession
pyspark_session = SparkSession(pyspark_context)
from pyspark.sql import SparkSession
```

Listing 4-11 changes the pandas dataframe created earlier in this chapter to a PySpark dataframe using the createDataFrame() method.

***Listing 4-11.*** Change the Pandas Dataframe to a PySpark Dataframe

```
pyspark_initial_data = pyspark_session.createDataFrame(initial_data)
```

Listing 4-12 creates a list for independent features and a string for the dependent feature. It the converts the data using the VectorAssembler() method for modeling with the PySpark framework.

***Listing 4-12.*** Transform the Data

```
x_list = list(initial_data.iloc[::, 0:9].columns)
from pyspark.ml.feature import VectorAssembler
pyspark_data_columns = x_list
pyspark_vector_assembler = VectorAssembler(inputCols=pyspark_data_columns,
outputCol="variables")
pyspark_data = pyspark_vector_assembler.transform(pyspark_initial_data)
```

Listing 4-13 executes the PySpark accelerated failure time method.

**Listing *4-13*.** Execute the PySpark Accelerated Failure Time Method

```
from pyspark.ml.regression import AFTSurvivalRegression
pyspark_accelerated_failure_method = AFTSurvivalRegression(censorCol=pyspa
rk_data.columns[1], labelCol=pyspark_data.columns[0],featuresCol="variables",)
pyspark_accelerated_failure_method_fitted = pyspark_accelerated_failure_
method.fit(pyspark_data)
```

Listing 4-14 computes the PySpark accelerated failure time method's predictions.

**Listing *4-14*.** Compute the PySpark Accelerated Failure Time Method's Predictions

```
pyspark_accelerated_failure_method_fitted.transform(pyspark_data).
select(pyspark_data.columns[1],"prediction")
pyspark_yhat.show()
```

```
+------+------------------+
|arrest|        prediction|
+------+------------------+
|     1|18.883982665910125|
|     1| 16.88228128814963|
|     1|22.631360777172517|
|     0|373.13041474613107|
|     0| 377.2238319806288|
|     0| 375.8326538406928|
|     1|  20.9780526816987|
|     0| 374.6420738270714|
|     0| 379.7483494080467|
|     0| 376.1601473382181|
|     0| 377.1412349521787|
|     0| 373.7536844216336|
|     1| 36.36443059383637|
|     0|374.14261327949384|
|     1| 22.98494042401171|
|     1| 50.61463874375869|
```

```
|     1| 25.56399364288275|
|     0|379.61997114629696|
|     0| 384.3322960430372|
|     0|376.37634062210844|
+------+------------------+
```

Listing 4-15 computes the PySpark accelerated failure time method's coefficients.

***Listing 4-15.*** Compute the PySpark Accelerated Failure Time Method's Coefficients

```
pyspark_accelerated_failure_method_fitted.coefficients
```

```
DenseVector([0.0388, -1.7679, -0.0162, -0.0003, 0.0098, -0.0086, -0.0026, 0.0115, 0.0003])
```

# Conclusion

This chapter executed two key machine learning frameworks (Lifeline and PySpark) to model censored data with the Cox Proportional Hazards and accelerated failure time methods.

# CHAPTER 5

# Nonlinear Modeling With Scikit-Learn, PySpark, and H2O

This chapter executes and appraises a nonlinear method for binary classification (called *logistic regression*) using a diverse set of comprehensive Python frameworks (i.e., Scikit-Learn, Spark MLlib, and H2O). To begin, it clarifies the underlying concept behind the sigmoid function.

## Exploring the Logistic Regression Method

The logistic regression method unanimously accepts values and then models them by executing a function (*sigmoid*) to anticipate values of a categorical output feature. Equation 5-1 defines the sigmoid function, which applies to logistic regression (also see Figure 5-1).

$$S(x) = \frac{L}{1 - e^{-x}} = \frac{e^x}{e^x + 1} \qquad \text{(Equation 5-1)}$$

Both Equation 5-1 and Figure 5-1 suggest that the function produces binary output values.

© Tshepo Chris Nokeri 2022
T. C. Nokeri, *Data Science Solutions with Python*, https://doi.org/10.1007/978-1-4842-7762-1_5

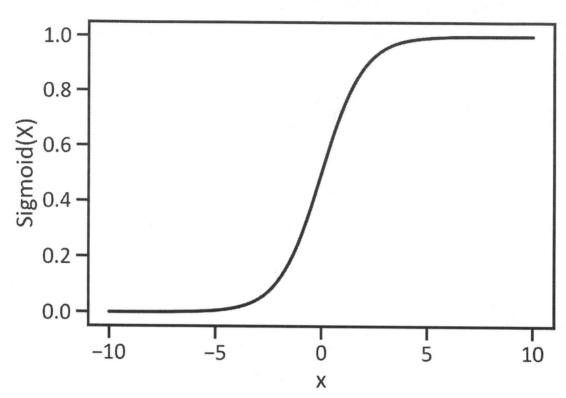

**Figure 5-1.** *Sigmoid function*

Listing 5-1 attains the necessary data from a Microsoft CSV file using the pandas framework.

**Listing 5-1.** Attain the Data

```
import pandas as pd
df = pd.read_csv(r"C:\Users\i5 lenov\Downloads\banking.csv")
```

Listing 5-2 stipulates the names of columns to drop and then executes the drop() method. It stipulates axes as columns in order to drop the unnecessary columns in the data.

**Listing 5-2.** Drop Unnecessary Features in the Data

```
drop_column_names = df.columns[[8, 9, 10]]
initial_data = df.drop(drop_column_names, axis="columns")
```

Listing 5-3 attains dummy values for categorical features in the data.

***Listing 5-3.*** Attain Dummy Features

```
initial_data.iloc[::, 1] = pd.get_dummies(initial_data.iloc[::, 1])
initial_data.iloc[::, 2] = pd.get_dummies(initial_data.iloc[::, 2])
initial_data.iloc[::, 3] = pd.get_dummies(initial_data.iloc[::, 3])
initial_data.iloc[::, 4] = pd.get_dummies(initial_data.iloc[::, 4])
initial_data.iloc[::, 5] = pd.get_dummies(initial_data.iloc[::, 5])
initial_data.iloc[::, 6] = pd.get_dummies(initial_data.iloc[::, 6])
initial_data.iloc[::, 7] = pd.get_dummies(initial_data.iloc[::, 7])
initial_data.iloc[::, 11] = pd.get_dummies(initial_data.iloc[::, 11])
```

Listing 5-4 drops the null values.

***Listing 5-4.*** Drop Null Values

```
initial_data = initial_data.dropna()
```

# Scikit-Learn in Action

This section executes and assesses the logistic regression method with the Scikit-Learn framework. Listing 5-5 outlines the independent and dependent features.

***Listing 5-5.*** Outline the Features

```
import numpy as np
x = np.array(initial_data.iloc[::, 0:17])
y = np.array(initial_data.iloc[::,-1])
```

Listing 5-6 randomly divides the dataframe.

***Listing 5-6.*** Randomly Divide the Dataframe

```
from sklearn.model_selection import train_test_split
x_train, x_test, y_train, y_test = train_test_split(x, y, test_size=0.2,
random_state=0)
```

Listing 5-7 scales the independent features.

***Listing 5-7.*** Scale Independent Features

```
from sklearn.preprocessing import StandardScaler
sk_standard_scaler = StandardScaler()
sk_standard_scaled_x_train = sk_standard_scaler.fit_transform(x_train)
sk_standard_scaled_x_test = sk_standard_scaler.transform(x_test)
```

Listing 5-8 executes the Scikit-Learn logistic regression method.

***Listing 5-8.*** Execute the Scikit-Learn Logistic Regression Method

```
from sklearn.linear_model import LogisticRegression
sk_logistic_regression_method = LogisticRegression()
sk_logistic_regression_method.fit(sk_standard_scaled_x_train,  y_train)
```

Listing 5-9 determines the best hyperparameters for the Scikit-Learn logistic regression method.

***Listing 5-9.*** Determine the Best Hyperparameters for the Scikit-Learn Logistic Regression Method

```
from sklearn.model_selection import GridSearchCV
sk_logistic_regression_method_param = {"penalty":("l1","l2")}
sk_logistic_regression_method_param_mod  = GridSearchCV(estimator=sk_
logistic_regression_method, param_grid=sk_logistic_regression_method_param,
n_jobs=-1)
sk_logistic_regression_method_param_mod.fit(sk_standard_scaled_x_train, y_train)
print("Best logistic regression score: ", sk_logistic_regression_method_
param_mod.best_score_)
print("Best logistic regression parameter: ", sk_logistic_regression_
method_param_mod.best_params_)
Best logistic regression score:  0.8986039453717755
Best logistic regression parameter:  {'penalty': 'l2'}
```

Listing 5-10 executes the logistic regression method with the Scikit-Learn framework.

***Listing 5-10.*** Execute the Scikit-Learn Logistic Regression Method

```
sk_logistic_regression_method = LogisticRegression(penalty="l2")
sk_logistic_regression_method.fit(sk_standard_scaled_x_train, y_train)
```

Listing 5-11 computes the logistic regression method's intercept.

***Listing 5-11.*** Compute the Logistic Regression Method's Intercept

```
print(sk_logistic_regression_method.intercept_)
[-2.4596243]
```

Listing 5-12 computes the coefficients.

***Listing 5-12.*** Compute the Logistic Regression Method's Coefficients

```
print(sk_logistic_regression_method.coef_)
[[ 0.03374725  0.04330667 -0.01305369 -0.02709009  0.13508899  0.01735913
   0.00816758  0.42948983 -0.12670658 -0.25784955 -0.04025993 -0.14622466
  -1.14143485  0.70803518  0.23256046 -0.02295578 -0.02857435]]
```

Listing 5-13 computes the Scikit-Learn logistic regression method's confusion matrix, which includes two forms of errors—false positives and false negatives and true positives and true negatives (see Table 5-1).

***Listing 5-13.*** Compute the Scikit-Learn Logistic Regression Method's Confusion Matrix

```
from sklearn import metrics
sk_logistic_regression_method_assessment_1 = pd.DataFrame(metrics.
confusion_matrix(y_test, sk_yhat), index=["Actual: Deposit","Actual: No
deposit"], columns=("Predicted: deposit","Predicted: No deposit"))
print(sk_logistic_regression_method_assessment_1)
```

***Table 5-1.*** *Scikit-Learn Logistic Regression Method's Confusion Matrix*

|  | **Predicted: Deposit** | **Predicted: No Deposit** |
|---|---|---|
| Actual: Deposit | 7230 | 95 |
| Actual: No Deposit | 711 | 202 |

Listing 5-14 computes the appropriate classification report (see Table 5-2).

**Listing 5-14.** Compute the Scikit-Learn Logistic Regression Method's Classification Report

```
sk_logistic_regression_method_assessment_2 = pd.DataFrame(metrics.
classification_report(y_test, sk_yhat, output_dict=True)).transpose()
print(sk_logistic_regression_method_assessment_2)
```

**Table 5-2.** *Scikit-Learn Logistic Regression Method's Classification Report*

|  | **Precision** | **Recall** | **F1-score** | **Support** |
|---|---|---|---|---|
| 0 | 0.910465 | 0.987031 | 0.947203 | 7325.000000 |
| 1 | 0.680135 | 0.221249 | 0.333884 | 913.000000 |
| Accuracy | 0.902161 | 0.902161 | 0.902161 | 0.902161 |
| Macro Avg | 0.795300 | 0.604140 | 0.640544 | 8238.000000 |
| Weighted Avg | 0.884938 | 0.902161 | 0.879230 | 8238.000000 |

Listing 5-15 arranges the Scikit-Learn logistic regression method's receiver operating characteristics curve. The goal is to condense the arrangement of the true positive rate (the proclivity of the method to correctly differentiate positive classes) and the false positive rate (the proclivity of the method to correctly differentiate negative classes). See Figure 5-2.

**Listing 5-15.** Receiver Operating Characteristics Curve for the Scikit-Learn Logistic Regression Method

```
sk_yhat_proba = sk_logistic_regression_method.predict_proba(sk_standard_
scaled_x_test)[::,1]
fpr_sk_logistic_regression_method, tprr_sk_logistic_regression_method, _ =
metrics.roc_curve(y_test, sk_yhat_proba)
area_under_curve_sk_logistic_regression_method = metrics.roc_auc_score(y_
test, sk_yhat_proba)
plt.plot(fpr_sk_logistic_regression_method, tprr_sk_logistic_regression_
method, label="AUC= "+ str(area_under_curve_sk_logistic_regression_method))
plt.xlabel("False Positive Rate (FPR)")
plt.ylabel("True Positive Rate (TPR)")
plt.legend(loc="best")
plt.show()
```

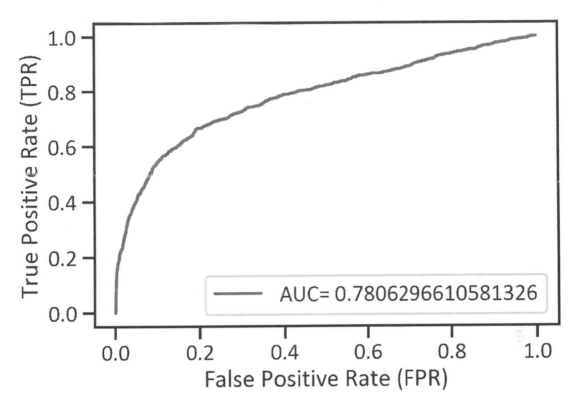

***Figure 5-2.*** *Receiver operating characteristics curve for the Scikit-Learn logistic regression method*

Listing 5-16 arranges the Scikit-Learn logistic regression method's precision-recall curve to condense the arrangement of the precision and recall (see Figure 5-3).

***Listing 5-16.*** Precision-Recall Curve for the Scikit-Learn Logistic Regression Method

```
p_sk_logistic_regression_method, r__sk_logistic_regression_method, _ =
metrics.precision_recall_curve(y_test, sk_yhat)
weighted_ps_sk_logistic_regression_method = metrics.roc_auc_score(y_test,
sk_yhat)
plt.plot(p_sk_logistic_regression_method, r__sk_logistic_regression_method,
        label="WPR= " +str(weighted_ps_sk_logistic_regression_method))
plt.xlabel("Recall")
plt.ylabel("Precision")
plt.legend(loc="best")
plt.show()
```

45

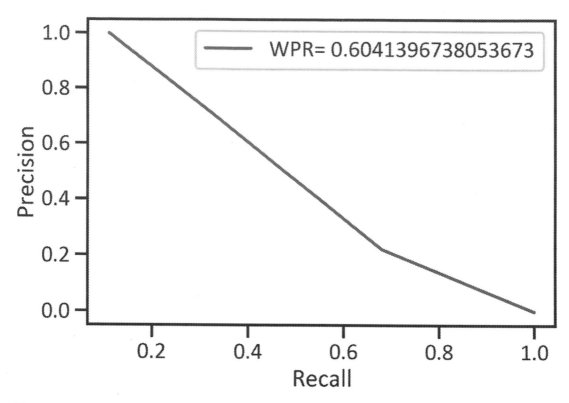

**Figure 5-3.** *Precision-recall curve for the Scikit-Learn logistic regression method*

Listing 5-17 arranges the learning curve for the Scikit-Learn logistic regression method to disclose the variations in weighted training and cross-validation accuracy (see Figure 5-4).

**Listing 5-17.** Learning Curve for the Logistic Regression Method Executed by Scikit-Learn

```
from sklearn.model_selection import learning_curve
train_port_sk_logistic_regression_method, trainscoresk_logistic_regression_
method, testscoresk_logistic_regression_method = learning_curve(sk_
logistic_regression_method, x, y,
                    cv=3, n_jobs=-5, train_sizes=np.linspace(0.1,1.0,50))
trainscoresk_logistic_regression_method_mean = np.mean(trainscoresk_
logistic_regression_method, axis=1)
```

```
testscoresk_logistic_regression_method_mean = np.mean(testscoresk_logistic_
regression_method, axis=1)
plt.plot(train_port_sk_logistic_regression_method, trainscoresk_logistic_
regression_method_mean, label="Weighted training accuracy")
plt.plot(train_port_sk_logistic_regression_method, testscoresk_logistic_
regression_method_mean, label="Weighted cv accuracy Score")
plt.xlabel("Training values")
plt.ylabel("Weighted accuracy score")
plt.legend(loc="best")
plt.show()
```

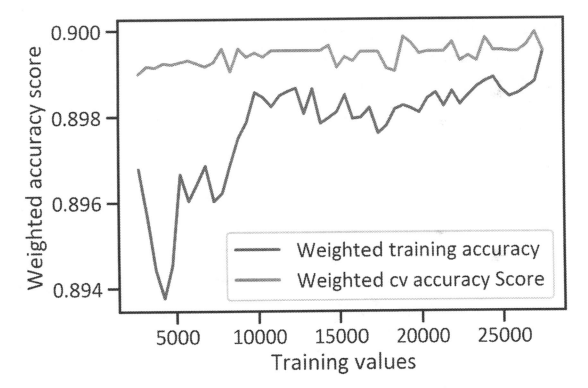

***Figure 5-4.*** *Learning curve for the logistic regression method executed by Scikit-Learn*

# PySpark in Action

This section executes and assesses the logistic regression method with the PySpark framework.

Listing 5-18 prepares the PySpark framework using the `findspark` framework.

***Listing 5-18.*** Prepare the PySpark Framework

```
import findspark as initiate_pyspark
initiate_pyspark.init("filepath\spark-3.0.0-bin-hadoop2.7")
```

Listing 5-19 stipulates the PySpark app using the `SparkConf()` method.

***Listing 5-19.*** Stipulate the PySpark App

```
from pyspark import SparkConf
pyspark_configuration = SparkConf().setAppName("pyspark_logistic_
regression_method").setMaster("local")
```

Listing 5-20 prepares the PySpark session using the `SparkSession()` method.

***Listing 5-20.*** Prepare the Spark Session

```
from pyspark.sql import SparkSession
pyspark_session = SparkSession(pyspark_context)
```

Listing 5-21 changes the `pandas` dataframe created earlier in this chapter to a PySpark dataframe using the `createDataFrame()` method.

***Listing 5-21.*** Change the Pandas Dataframe to a PySpark Dataframe

```
pyspark_initial_data = pyspark_session.createDataFrame(initial_data)
```

Listing 5-22 creates a list of independent features and a string for the dependent feature. It then converts the data using the `VectorAssembler()` method for modeling with the PySpark framework.

***Listing 5-22.*** Transform the Data

```
x_list = list(initial_data.iloc[::, 0:17].columns)
y_list = str(initial_data.columns[-1])
from pyspark.ml.feature import VectorAssembler
```

```
pyspark_data_columns = x_list
pyspark_vector_assembler = VectorAssembler(inputCols=pyspark_data_columns,
outputCol="features")
pyspark_data = pyspark_vector_assembler.transform(pyspark_initial_data)
```

Listing 5-23 divides the data using the randomSplit() method.

***Listing 5-23.*** Divide the Dataframe

```
(pyspark_training_data, pyspark_test_data) = pyspark_data.randomSplit([.8,.2])
```

Listing 5-24 executes the PySpark logistic regression method.

***Listing 5-24.*** Execute the PySpark Logistic Regression Method

```
from pyspark.ml.classification import LogisticRegression
pyspark_logistic_regression_method = LogisticRegression(labelCol = y_list,
featuresCol = "features")
pyspark_logistic_regression_method_fitted = pyspark_logistic_regression_
method.fit(pyspark_training_data)
```

Listing 5-25 computes the PySpark logistic regression method's predictions.

***Listing 5-25.*** Logistic Regression Method Predictions (Method Executed with PySpark Framework)

```
pyspark_yhat = pyspark_logistic_regression_method_fitted.transform(pyspark_
test_data)
```

Listing 5-26 arranges the PySpark logistic regression method's receiver operating characteristics curve to condense the arrangement of the precision and recall (see Figure 5-5).

***Listing 5-26.*** Receiver Operating Characteristics Curve for the PySpark Logistic Regression Method

```
pyspark_logistic_regression_method_assessment = pyspark_logistic_
regression_method_fitted.summary
pyspark_logistic_regression_method_roc = pyspark_logistic_regression_
method_assessment.roc.toPandas()
pyspark_logistic_regression_method_auroc = pyspark_logistic_regression_
method_assessment.areaUnderROC
```

```
plt.plot(pyspark_logistic_regression_method_roc["FPR"], pyspark_logistic_
regression_method_roc["TPR"],
          label="AUC= "+str(pyspark_logistic_regression_method_auroc))
plt.xlabel("False Positive Rate (FPR)")
plt.ylabel("True Positive Rate (TPR)")
plt.legend(loc=4)
plt.show()
```

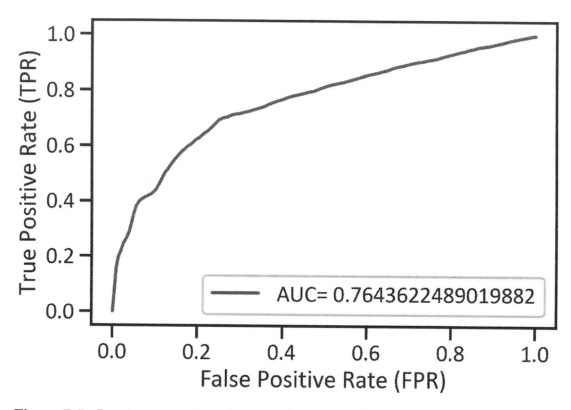

***Figure 5-5.*** *Receiver operating characteristics curve for the PySpark logistic regression method*

Listing 5-27 arranges the PySpark logistic regression method's precision-recall curve to condense the arrangement of the precision and recall (Figure 5-6).

**Listing 5-27.** Precision-Recall Curve for the PySpark Logistic Regression Method

```
pyspark_logistic_regression_method_assessment = pyspark_logistic_
regression_method_fitted.summary
pyspark_logistic_regression_method_assessment_pr = pyspark_logistic_
regression_method_assessment.pr.toPandas()
pyspark_logistic_regression_method_assessment_wpr = pyspark_logistic_
regression_method_assessment.weightedPrecision
plt.plot(pyspark_logistic_regression_method_assessment_pr["precision"],
        pyspark_logistic_regression_method_assessment_pr["recall"],
        label="WPR: "+str(pyspark_logistic_regression_method_assessment_
        wpr))
plt.xlabel("Precision")
plt.ylabel("Recall")
plt.legend(loc="best")
plt.show()
```

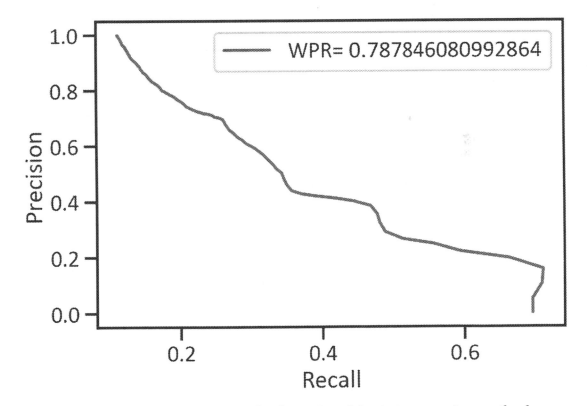

**Figure 5-6.** *Precision-recall curve for the PySpark logistic regression method*

# H2O in Action

This section executes and assesses the logistic regression method using the H2O framework. Listing 5-28 prepares the H2O framework.

***Listing 5-28.*** Prepare the H2O Framework

```
import h2o as initialize_h2o
initialize_h2o.init()
```

Listing 5-29 changes the pandas dataframe to the H2O dataframe.

***Listing 5-29.*** Change the Pandas Dataframe to H2O Dataframe

```
h2o_data = initialize_h2o.H2OFrame(initial_data)
```

Listing 5-30 outlines the independent and dependent features.

***Listing 5-30.*** Outline the Features

```
x_list = list(initial_data.iloc[::, 0:17].columns)
y_list = str(initial_data.columns[-1])
y = y_list
x = h2o_data.col_names
x.remove(y_list)
```

Listing 5-31 randomly divides the data.

***Listing 5-31.*** Randomly Divide the Dataframe

```
h2o_training_data, h2o_validation_data, h2o_test_data = h2o_data.split_
frame(ratios=[.8,.1])
```

Listing 5-32 executes the H2O logistic regression method.

***Listing 5-32.*** Execute the H2O Logistic Regression Method

```
from h2o.estimators.glm import H2OGeneralizedLinearEstimator
h2o_logistic_regression_method = H2OGeneralizedLinearEstimator(family=
"binomial")
h2o_logistic_regression_method.train(x = x, y = y, training_frame =
h2o_training_data, validation_frame = h2o_validation_data)
```

Listing 5-33 computes the H2O logistic regression method's predictions.

***Listing 5-33.*** Compute the H2O Logistic Regression Method's Predictions

```
h2o_yhat = h2o_logistic_regression_method.predict(h2o_test_data)
```

Listing 5-34 computes the H2O logistic regression method's predictions (see Figure 5-7).

***Listing 5-34.*** Compute the H2O Logistic Regression Method's Standardized Coefficients

```
h2o_logistic_regression_method_std_coefficients = h2o_logistic_regression_
method.std_coef_plot()
h2o_logistic_regression_method_std_coefficients
```

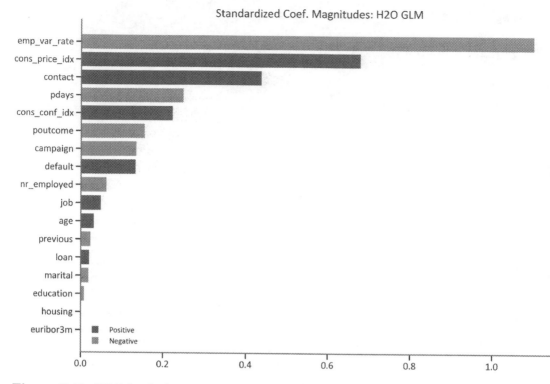

***Figure 5-7.*** *H2O logistic regression method's standardized coefficients*

Listing 5-35 computes the H2O logistic regression method's partial dependence (see Figure 5-8).

***Listing 5-35.*** Compute the H2O Logistic Regression Method's Partial Dependency

```
h2o_logistic_regression_method_dependency_plot = h2o_logistic_regression_
method.partial_plot(data = h2o_data, cols = list(initial_data.columns
[[0, 17]]), server=False, plot = True)
h2o_logistic_regression_method_dependency_plot
```

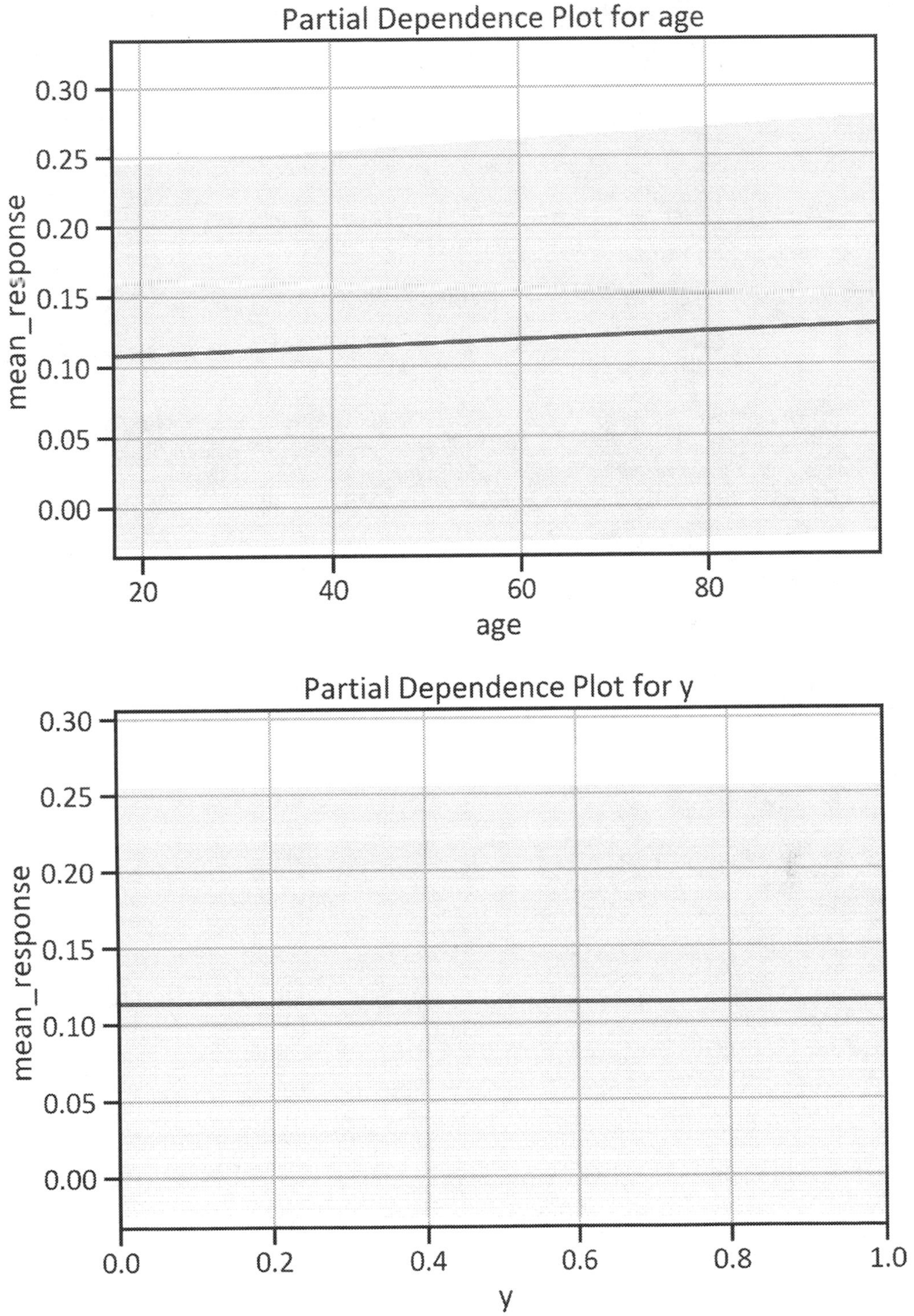

*Figure 5-8.* *H2O logistic regression method's standardized coefficients*

Listing 5-36 arranges the features that are most important to the H2O logistic regression method in ascending order (see Figure 5-9).

**Listing 5-36.** Compute the H2O Logistic Regression Method's Variance Importance

```
h2o_logistic_regression_method_feature_importance = h2o_logistic_
regression_method.varimp_plot()
h2o_logistic_regression_method_feature_importance
```

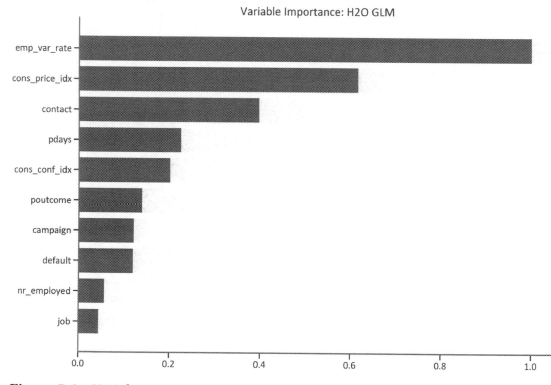

**Figure 5-9.** *H2O logistic regression method's variance importance*

Listing 5-37 arranges the H2O logistic regression method's receiver operating characteristics curve to condense the arrangement of the true positive rate and false positive rate (see Figure 5-10).

***Listing 5-37.*** Receiver Operating Characteristics Curve for the H2O Logistic Regression Method

```
h2o_logistic_regression_method_assessment = h2o_logistic_regression_method.
model_performance()
```

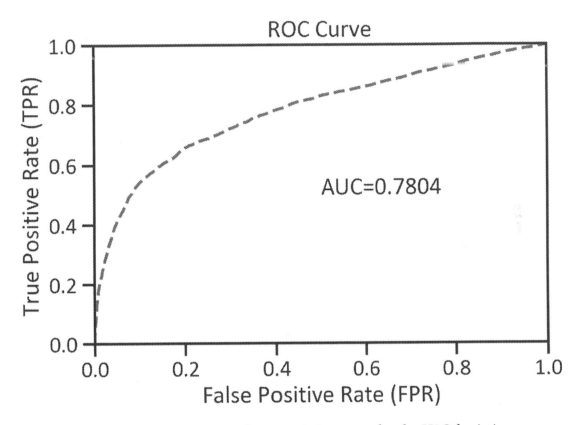

***Figure 5-10.*** *Receiver operating characteristics curve for the H2O logistic regression method*

# Conclusion

This chapter executed three key machine learning frameworks (Scikit-Learn, PySpark, and H2O) to model data and spawn a categorical output feature with two classes using the logistic regression method.

# CHAPTER 6

# Tree Modeling and Gradient Boosting with Scikit-Learn, XGBoost, PySpark, and H2O

This chapter executes and appraises a tree-based method (the decision tree method) and an ensemble method (the gradient boosting trees method) using a diverse set of comprehensive Python frameworks (i.e., Scikit-Learn, XGBoost, PySpark, and H2O). To begin, the chapter clarifies how decision trees compute the probabilities of classes.

## Decision Trees

The *decision tree* is an elementary, non-parametric method suitable for linear and nonlinear modeling. It executes decision rules and develops a tree-like structure that divides values into varying groups with the least depth (see a Figure 6-1).

© Tshepo Chris Nokeri 2022
T. C. Nokeri, *Data Science Solutions with Python*, https://doi.org/10.1007/978-1-4842-7762-1_6

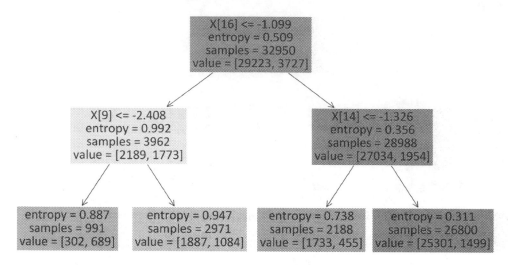

*Figure 6-1.* *Decision tree*

# Preprocessing Features

This chapter manipulates the data from Chapter 5, so it does not sequentially cover the preprocessing tasks. Listing 6-1 executes all the preprocessing tasks.

*Listing 6-1.* Preprocess Features

```
from sklearn.tree import DecisionTreeClassifier
import numpy as np
import pandas as pd
from sklearn.preprocessing import StandardScaler
from sklearn.model_selection import train_test_split
df = pd.read_csv(r"filepath\banking.csv")
drop_column_names = df.columns[[8, 9, 10]]
initial_data = df.drop(drop_column_names, axis="columns")
initial_data.iloc[::, 1] = pd.get_dummies(initial_data.iloc[::, 1])
initial_data.iloc[::, 2] = pd.get_dummies(initial_data.iloc[::, 2])
initial_data.iloc[::, 3] = pd.get_dummies(initial_data.iloc[::, 3])
initial_data.iloc[::, 4] = pd.get_dummies(initial_data.iloc[::, 4])
initial_data.iloc[::, 5] = pd.get_dummies(initial_data.iloc[::, 5])
initial_data.iloc[::, 6] = pd.get_dummies(initial_data.iloc[::, 6])
initial_data.iloc[::, 7] = pd.get_dummies(initial_data.iloc[::, 7])
```

```
initial_data.iloc[::, 11] = pd.get_dummies(initial_data.iloc[::, 11])
initial_data = initial_data.dropna()
x = np.array(initial_data.iloc[::, 0:17])
y = np.array(initial_data.iloc[::, -1])
x_train, x_test, y_train, y_test = train_test_split(x, y, test_size=0.2,
random_state=0)
sk_standard_scaler = StandardScaler()
sk_standard_scaled_x_train = sk_standard_scaler.fit_transform(x_train)
sk_standard_scaled_x_test = sk_standard_scaler.transform(x_test)
```

## Scikit-Learn in Action

This section executes and assesses the decision tree method with the Scikit-Learn framework. Listing 6-2 executes the decision tree method with the Scikit-Learn framework.

***Listing 6-2.*** Execute the Scikit-Learn Decision Tree Method

```
from sklearn.tree import DecisionTreeClassifier
sk_decision_tree_method = DecisionTreeClassifier()
sk_decision_tree_method.fit(sk_standard_scaled_x_train, y_train)
```

Listing 6-3 determines the best hyperparameters for the Scikit-Learn decision tree method.

***Listing 6-3.*** Determine the Best Hyperparameters for the Scikit-Learn Decision Tree Method

```
from sklearn.model_selection import GridSearchCV
sk_decision_tree_method_parameters = {"criterion":("gini","entropy"),
"max_depth":[1, 2, 3, 4, 5, 6]}
sk_decision_tree_method_g_search = GridSearchCV(estimator = sk_decision_
tree_method, param_grid = sk_decision_tree_method_parameters)
sk_decision_tree_method_g_search.fit(sk_standard_scaled_x_train, y_train)
print("Best decision tree regression score: ", sk_decision_tree_method_g_
search.best_score_)
print("Best decision tree parameter: ", sk_decision_tree_method_g_search.
best_params_)
```

```
Best decision tree regression score:  0.900030349013657
Best decision tree parameter:  {'criterion': 'entropy', 'max_depth': 5}
```

Listing 6-4 executes the decision tree method using the Scikit-Learn framework.

***Listing 6-4.*** Execute the Scikit-Learn Decision Tree Method

```
sk_decision_tree_method = DecisionTreeClassifier(criterion = "entropy",
max_depth = 5)
sk_decision_tree_method.fit(sk_standard_scaled_x_train, y_train)
```

Listing 6-5 arranges the Scikit-Learn decision tree method's classification report (see Table 6-1).

***Listing 6-5.*** Arrange the Scikit-Learn Decision Tree Method's Classification Report

```
from sklearn import metrics
sk_yhat = sk_decision_tree_method.predict(sk_standard_scaled_x_test)
sk_decision_tree_method_assessment_2 = pd.DataFrame(metrics.classification_
report(y_test, sk_yhat, output_dict=True)).transpose()
sk_decision_tree_method_assessment_2
```

***Table 6-1.*** *Scikit-Learn Decision Tree Method's Classification Report*

|  | Precision | Recall | F1-score | Support |
|---|---|---|---|---|
| 0 | 0.914696 | 0.982253 | 0.947271 | 7325.000000 |
| 1 | 0.650538 | 0.265060 | 0.376654 | 913.000000 |
| Accuracy | 0.902768 | 0.902768 | 0.902768 | 0.902768 |
| Macro Avg | 0.782617 | 0.623656 | 0.661963 | 8238.000000 |
| Weighted Avg | 0.885420 | 0.902768 | 0.884031 | 8238.000000 |

Listing 6-6 arranges the decision tree method's receiver operating characteristics curve to condense the arrangement of the true positive rate and the false positive rate (see Figure 6-2).

***Listing 6-6.*** Receiver Operating Characteristics Curve for the Decision Tree
Method (Executed by the Scikit-Learn Framework)

```
import matplotlib.pyplot as plt
%matplotlib inline
sk_yhat_proba = sk_decision_tree_method.predict_proba(sk_standard_scaled_
x_test)[::,1]
fpr_sk_decision_tree_method, tprr_sk_decision_tree_method, _ = metrics.
roc_curve(y_test, sk_yhat_proba)
area_under_curve_sk_decision_tree_method = metrics.roc_auc_
score(y_test, sk_yhat_proba)
plt.plot(fpr_sk_decision_tree_method, tprr_sk_decision_tree_method,
label="AUC= "+ str(area_under_curve_sk_decision_tree_method))
plt.xlabel("False Positive Rate (FPR)")
plt.ylabel("True Positive Rate (TPR)")
plt.legend(loc="best")
plt.show()
```

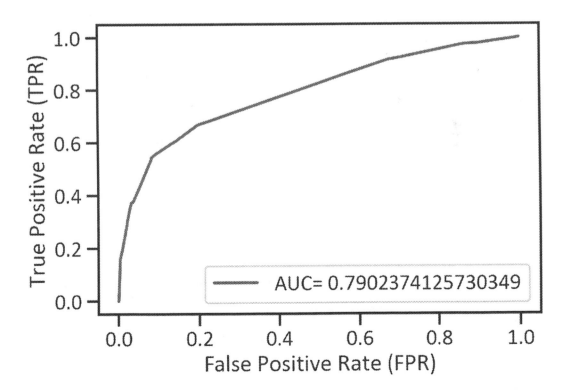

***Figure 6-2.*** *Receiver operating characteristics curve for the Scikit-Learn decision*

Listing 6-7 arranges the Scikit-Learn decision tree method's receiver operating characteristics curve to condense the arrangement of the precision and recall (see Figure 6-3).

***Listing 6-7.*** Precision-Recall Curve for the Scikit-Learn Decision Tree Method

```
p_sk_decision_tree_method, r__sk_decision_tree_method, _ = metrics.
precision_recall_curve(y_test, sk_yhat)
weighted_ps_sk_decision_tree_method = metrics.roc_auc_score(y_test, sk_yhat)
plt.plot(p_sk_decision_tree_method, r__sk_decision_tree_method,
         label="WPR= " +str(weighted_ps_sk_decision_tree_method))
plt.xlabel("Recall")
plt.ylabel("Precision")
plt.legend(loc="best")
plt.show()
```

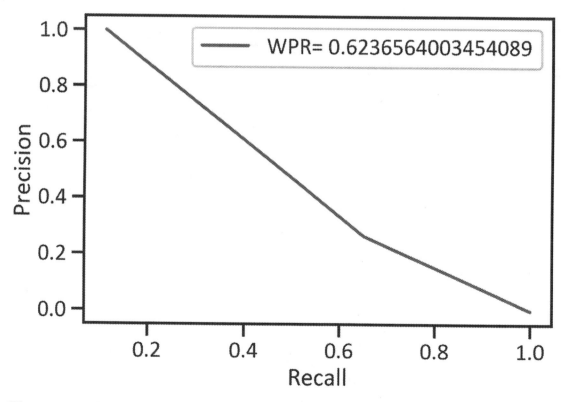

***Figure 6-3.*** *Precision-recall curve for the Scikit-Learn framework decision tree method*

Listing 6-8 arranges the Scikit-Learn decision tree method's learning curve (see Figure 6-4).

***Listing 6-8.*** Learning Curve for the Decision Tree Method Executed by the Scikit-Learn Framework

```
from sklearn.model_selection import learning_curve
train_port_sk_decision_tree_method, trainscore_sk_decision_tree_method,
testscore_sk_decision_tree_method = learning_curve(sk_decision_tree_method,
                                x, y, cv=3, n_jobs=-5, train_sizes=np.
                                linspace(0.1,1.0,50))
trainscoresk_decision_tree_method_mean = np.mean(trainscore_sk_decision_
tree_method, axis=1)
testscoresk_decision_tree_method_mean = np.mean(testscore_sk_decision_tree_
method, axis=1)
plt.plot(train_port_sk_decision_tree_method, trainscoresk_decision_tree_
method_mean, label="Weighted training accuracy")
plt.plot(train_port_sk_decision_tree_method, testscoresk_decision_tree_
method_mean, label="Weighted cv accuracy Score")
plt.xlabel("Training values")
plt.ylabel("Weighted accuracy score")
plt.legend(loc="best")
plt.show()
```

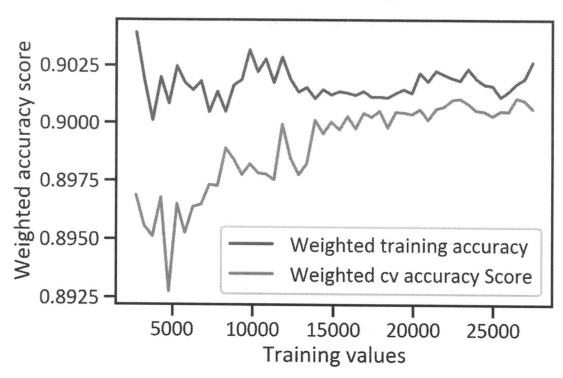

*Figure 6-4.* *Learning curve for the decision tree method executed by Scikit-Learn*

# Gradient Boosting

Gradient boosting methods inherit values of input features and then execute countless tree models to halt the loss function (i.e., the mean absolute error for linear modeling). It does this by assimilating weak models, then incrementally and iteratively models weighing data, diligently accompanied by an election of a weak model with the best performance.

# XGBoost in Action

This section executes and assesses the decision tree method with the XGBoost framework. Listing 6-9 executes the XGBoost gradient boosting method.

*Listing 6-9.* Execute the XGBoost Gradient Boosting Method

```
from xgboost import XGBClassifier
xgb_gradient_boosting_method = XGBClassifier()
xgb_gradient_boosting_method.fit(sk_standard_scaled_x_train, y_train)
```

Listing 6-10 arranges the XGBoost gradient boosting method's classification report (see Table 6-2).

***Listing 6-10.*** Arrange the XGBoost Gradient Boosting Method's Classification Report

```
sk_yhat_xgb_gradient_boosting_method = sk_yhat_xgb_gradient_boosting_method
= xgb_gradient_boosting_method.predict(sk_standard_scaled_x_test)
xgb_gradient_boosting_method_assessment_2 = pd.DataFrame(metrics.
classification_report(y_test, sk_yhat_xgb_gradient_boosting_method, output_
dict=True)).transpose()
print(xgb_gradient_boosting_method_assessment_2)
```

***Table 6-2.*** *XGBoost Gradient Boosting Method's Classification Report*

|  | Precision | Recall | F1-score | Support |
|---|---|---|---|---|
| 0 | 0.913410 | 0.976382 | 0.943847 | 7325.00000 |
| 1 | 0.575980 | 0.257393 | 0.355791 | 913.000000 |
| Accuracy | 0.896698 | 0.896698 | 0.896698 | 0.896698 |
| Macro Avg | 0.744695 | 0.616888 | 0.649819 | 8238.000000 |
| Weighted Avg | 0.876013 | 0.896698 | 0.878674 | 8238.000000 |

Listing 6-11 arranges the XGBoost gradient boosting method's receiver operating characteristics curve to condense the arrangement of the precision and recall (see Figure 6-5).

***Listing 6-11.*** Receiver Operating Characteristics Curve for the XGBoost Gradient Boosting Method

```
yhat_proba_xgb_gradient_boosting_method = xgb_gradient_boosting_method.
predict_proba(sk_standard_scaled_x_test)[::,1]
fpr_xgb_gradient_boosting_method, tprr_xgb_gradient_boosting_method, _ =
metrics.roc_curve(y_test, yhat_proba_xgb_gradient_boosting_method)
area_under_curve_xgb_gradient_boosting_method = metrics.roc_auc_score(y_
test, yhat_proba_xgb_gradient_boosting_method)
```

```
plt.plot(fpr_xgb_gradient_boosting_method, tprr_xgb_gradient_boosting_
method, label="AUC= "+ str(area_under_curve_xgb_gradient_boosting_method))
plt.xlabel("False Positive Rate (FPR)")
plt.ylabel("True Positive Rate (TPR)")
plt.legend(loc="best")
plt.show()
```

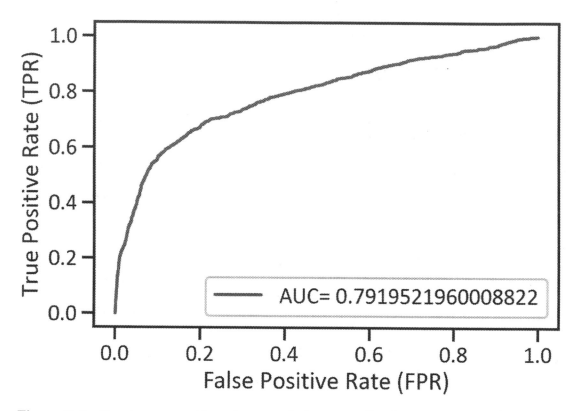

***Figure 6-5.*** *Receiver operating characteristics curve for the XGBoost gradient boosting method*

Listing 6-12 arranges the XGBoost gradient boosting method's precision-recall curve to condense the arrangement of the precision and recall (see Figure 6-6).

***Listing 6-12.*** Precision-Recall Curve for the Scikit-Learn Decision Tree Method

```
sk_yhat_xgb_gradient_boosting_method)
weighted_ps_xgb_gradient_boosting_method = metrics.roc_auc_score(y_test,
sk_yhat_xgb_gradient_boosting_method)
```

```
plt.plot(p_xgb_gradient_boosting_method, r__xgb_gradient_boosting_method,
        label="WPR= " +str(weighted_ps_xgb_gradient_boosting_method))
plt.xlabel("Recall")
plt.ylabel("Precision")
plt.legend(loc="best")
plt.show()
```

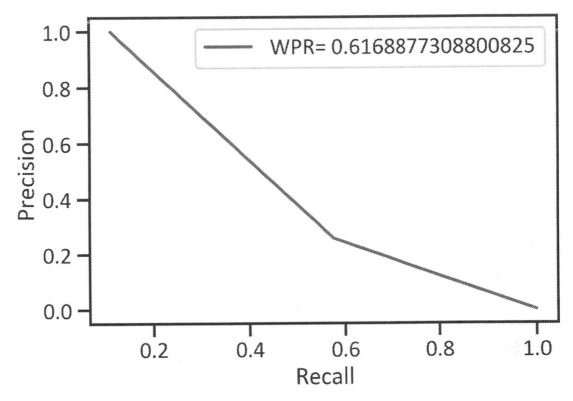

***Figure 6-6.*** *Precision-recall curve for the Scikit-Learn decision tree method*

## PySpark in Action

This section executes and assesses the gradient boosting method with the PySpark framework. Listing 6-13 prepares the PySpark framework using the findspark framework.

***Listing 6-13.*** Prepare the PySpark Framework

```
import findspark as initiate_pyspark
initiate_pyspark.init("filepath\spark-3.0.0-bin-hadoop2.7")
```

Listing 6-14 stipulates the PySpark app using the SparkConf() method.

***Listing 6-14.*** Stipulate the PySpark App

```
from pyspark import SparkConf
pyspark_configuration = SparkConf().setAppName("gradient_boosting_method").
setMaster("local")
```

Listing 6-15 prepares the PySpark session using the SparkSession() method.

***Listing 6-15.*** Prepare the Spark Session

```
from pyspark.sql import SparkSession
pyspark_session = SparkSession(pyspark_context)
```

Listing 6-16 changes the pandas dataframe created earlier in this chapter to a PySpark dataframe using the createDataFrame() method.

***Listing 6-16.*** Change the Pandas Dataframe to a PySpark Dataframe

```
pyspark_initial_data = pyspark_session.createDataFrame(initial_data)
```

Listing 6-17 creates a list for independent features and a string for the dependent feature. It then converts the data using the VectorAssembler() method to model with the PySpark framework.

***Listing 6-17.*** Transform the Data

```
x_list = list(initial_data.iloc[::, 0:17].columns)
y_list = str(initial_data.columns[-1])
from pyspark.ml.feature import VectorAssembler
pyspark_data_columns = x_list
pyspark_vector_assembler = VectorAssembler(inputCols=pyspark_data_columns,
outputCol="features")
pyspark_data = pyspark_vector_assembler.transform(pyspark_initial_data)
```

Listing 6-18 divides the data with the randomSplit() method.

***Listing 6-18.*** Divide the Dataframe

```
(pyspark_training_data, pyspark_test_data) = pyspark_data.
randomSplit([.8,.2])
```

Listing 6-19 executes the PySpark gradient boosting regression method.

***Listing 6-19.*** Execute the PySpark Gradient Boosting Method

```
from pyspark.ml.classification import GBTClassifier
pyspark_gradient_boosting_method = GBTClassifier(labelCol = y_list,
featuresCol = "features")
pyspark_gradient_boosting_method_fitted = pyspark_gradient_boosting_method.
fit(pyspark_training_data)
```

Listing 6-20 computes the gradient boosting regression method's predictions using the PySpark framework.

***Listing 6-20.*** Gradient Boosting Method Predictions (Method Executed with PySpark Framework)

```
pyspark_yhat = pyspark_gradient_boosting_method_fitted.transform(pyspark_
test_data)
```

# H2O in Action

This section executes and assesses the principal component method with the H2O framework. Listing 6-21 prepares the H2O framework.

***Listing 6-21.*** Prepare the H2O Framework

```
import h2o as initialize_h2o
initialize_h2o.init()
```

Listing 6-22 changes the pandas dataframe to an H2O dataframe.

***Listing 6-22.*** Change the Pandas Dataframe to an H2O Dataframe

```
h2o_data = initialize_h2o.H2OFrame(initial_data)
```

Listing 6-23 outlines the features.

**Listing 6-23.**  Outline the Features

```
x_list = list(initial_data.iloc[::, 0:17].columns)
y_list = str(initial_data.columns[-1])
y = y_list
x = h2o_data.col_names
x.remove(y_list)
```

Listing 6-24 randomly divides the dataframe.

**Listing 6-24.**  Randomly Divide the Dataframe

```
h2o_training_data, h2o_validation_data, h2o_test_data = h2o_data.split_
frame(ratios=[.8,.1])
```

Listing 6-25 executes the H2O gradient boosting method.

**Listing 6-25.**  Execute the H2O Gradient Boosting Method

```
from h2o.estimators import H2OGradientBoostingEstimator
h2o_gradient_boosting_method = H2OGradientBoostingEstimator(nfolds=3)
h2o_gradient_boosting_method.train(x = x, y = y, training_frame = h2o_
training_data, validation_frame = h2o_validation_data)
```

Listing 6-26 assesses the H2O gradient boosting method (see Table 6-3).

**Listing 6-26.**  Assess the H2O Gradient Boosting Method

```
h2o_gradient_boosting_method_history = h2o_gradient_boosting_method.
scoring_history()
print(h2o_gradient_boosting_method_history.head(5))
```

*Table 6-3. H2O Gradient Boosting Method Assessment*

| Timestamp | Duration | Number_ of_trees | Training_ rmse | Training_ mae | Training_ deviance | Validation_ rmse | Validation_ mae | Validation_ deviance | | |
|---|---|---|---|---|---|---|---|---|---|---|
| 0 | 2021- 08-24 04:04:30 | 10.384 sec | 0.0 | 0.317003 | 0.200981 | 0.100491 | 0.311883 | 0.197762 | 0.097271 |
| 1 | 2021- 08-24 04:04:30 | 10.496 sec | 1.0 | 0.309924 | 0.196310 | 0.096053 | 0.305219 | 0.193472 | 0.093159 |
| 2 | 2021- 08-24 04:04:30 | 10.586 sec | 2.0 | 0.304085 | 0.192122 | 0.092468 | 0.299762 | 0.189611 | 0.089857 |
| 3 | 2021- 08-24 04:04:30 | 10.659 sec | 3.0 | 0.299238 | 0.188351 | 0.089543 | 0.295379 | 0.186199 | 0.087249 |
| 4 | 2021- 08-24 04:04:30 | 10.745 sec | 4.0 | 0.295198 | 0.184939 | 0.087142 | 0.291736 | 0.183116 | 0.085110 |

73

# Conclusion

This chapter executed four key machine learning frameworks (Scikit-Learn, XGBoost, PySpark, and H2O) to model data and spawn a categorical output feature with two classes using the decision tree and gradient boosting methods.

# CHAPTER 7

# Neural Networks with Scikit-Learn, Keras, and H2O

This chapter executes and assesses nonlinear neural networks to address binary classification using a diverse set of comprehensive Python frameworks (i.e., Scikit-Learn, Keras, and H2O).

## Exploring Deep Learning

Deep learning methods transcend machine learning methods. They handle considerable complexity in models and methods by including extensive hidden layers. Machine learning methods typically resemble an input-output system, where an objective function accepts input values and transforms them with an objective function to produce values. Deep learning broadens the processing activities by including many layers that include nodes that repetitively and iteratively process data. The widespread recognition of deep learning relates to its customizability; it enable developers to customize methods based on preferences.

## Multilayer Perceptron Neural Network

As the name suggests, a multilayer perceptron neural network contains multiple layers. Moreover, the fundamental structure remains the same; there has to be one layer for receiving input values and one layer for generating output values. Layers that model the data to distribute differing weights and biases are treated as hidden layers.

© Tshepo Chris Nokeri 2022
T. C. Nokeri, *Data Science Solutions with Python*, https://doi.org/10.1007/978-1-4842-7762-1_7

Figure 7-1 shows the multilayer perceptron neural network that this chapter uses.

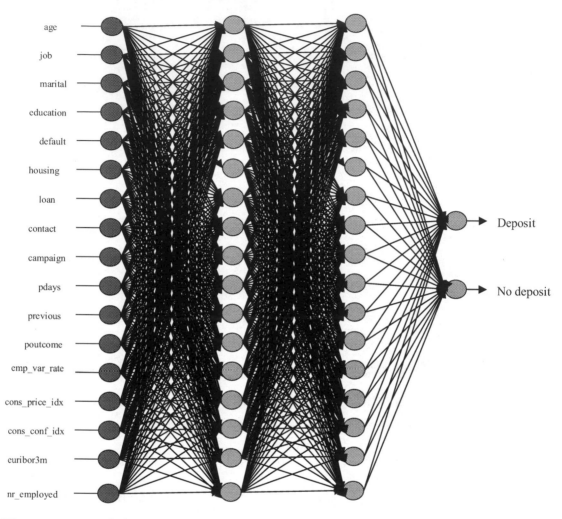

***Figure 7-1.***  *Multilayer perceptron neural network*

# Preprocessing Features

This chapter manipulates the data attained in Chapter 6, so it does not describe the preprocessing tasks in detail. Listing 7-1 executes all the preprocessing tasks.

*Listing 7-1.* Preprocess Features

```
import numpy as np
import pandas as pd
from sklearn.preprocessing import StandardScaler
from sklearn.model_selection import train_test_split
df = pd.read_csv(r"C:\Users\i5 lenov\Downloads\banking.csv")
drop_column_names = df.columns[[8, 9, 10]]
initial_data = df.drop(drop_column_names, axis="columns")
initial_data.iloc[::, 1] = pd.get_dummies(initial_data.iloc[::, 1])
initial_data.iloc[::, 2] = pd.get_dummies(initial_data.iloc[::, 2])
initial_data.iloc[::, 3] = pd.get_dummies(initial_data.iloc[::, 3])
initial_data.iloc[::, 4] = pd.get_dummies(initial_data.iloc[::, 4])
initial_data.iloc[::, 5] = pd.get_dummies(initial_data.iloc[::, 5])
initial_data.iloc[::, 6] = pd.get_dummies(initial_data.iloc[::, 6])
initial_data.iloc[::, 7] = pd.get_dummies(initial_data.iloc[::, 7])
initial_data.iloc[::, 11] = pd.get_dummies(initial_data.iloc[::, 11])
initial_data = initial_data.dropna()
x = np.array(initial_data.iloc[::,0:17])
y = np.array(initial_data.iloc[::,-1])
x_train, x_test, y_train, y_test = train_test_split(x, y, test_size=0.2,
random_state=0)
sk_standard_scaler = StandardScaler()
sk_standard_scaled_x_train = sk_standard_scaler.fit_transform(x_train)
sk_standard_scaled_x_test = sk_standard_scaler.transform(x_test)
```

# Scikit-Learn in Action

This section executes and assesses a multilayer perceptron method using the Scikit-Learn framework. Listing 7-2 executes the Scikit-Learn multilayer perceptron neural network.

*Listing 7-2.* Execute the Scikit-Learn Multilayer Perceptron Neural Network

```
from sklearn.neural_network import MLPClassifier
sk_multilayer_perceptron_net = MLPClassifier()
sk_multilayer_perceptron_net.fit(sk_standard_scaled_x_train, y_train)
```

Listing 7-3 arranges the Scikit-Learn multilayer perceptron neural network's classification report (see Table 7-1).

***Listing 7-3.*** Arrange the Scikit-Learn Multilayer Perceptron Neural Network's Classification Report

```
from sklearn import metrics
sk_yhat_multilayer_perceptron_net = sk_multilayer_perceptron_net.
predict(sk_standard_scaled_x_test)
sk_multilayer_perceptron_net_assessment = pd.DataFrame(metrics.
classification_report(y_test, sk_yhat_multilayer_perceptron_net,
                                        output_dict=True)).transpose()
print(sk_multilayer_perceptron_net_assessment)
```

***Table 7-1.*** *Scikit-Learn Multilayer Perceptron Neural Network's Classification Report*

|              | Precision | Recall   | F1-score | Support     |
|--------------|-----------|----------|----------|-------------|
| 0            | 0.917062  | 0.976655 | 0.945921 | 7325.000000 |
| 1            | 0.608696  | 0.291347 | 0.394074 | 913.000000  |
| Accuracy     | 0.900704  | 0.900704 | 0.900704 | 0.900704    |
| Macro Avg    | 0.762879  | 0.634001 | 0.669998 | 8238.000000 |
| Weighted Avg | 0.882886  | 0.900704 | 0.884761 | 8238.000000 |

Listing 7-4 arranges the multilayer perceptron neural network's receiver operating characteristics curve to condense the arrangement of the true positive rate and the false positive rate (see Figure 7-2).

***Listing 7-4.*** Arrange a Receiver Operating Characteristics Curve for the Multilayer Perceptron Network (Executed with the Scikit-Learn Framework)

```
import matplotlib.pyplot as plt
%matplotlib inline
yhat_proba_sk_multilayer_perceptron_net = sk_multilayer_perceptron_net.
predict_proba(sk_standard_scaled_x_test)[::,1]
```

```
fpr_sk_multilayer_perceptron_net, tprr_sk_multilayer_perceptron_net, _ =
metrics.roc_curve(y_test, yhat_proba_sk_multilayer_perceptron_net)
area_under_curve_sk_multilayer_perceptron_net = metrics.roc_auc_score(y_
test, yhat_proba_sk_multilayer_perceptron_net)
plt.plot(fpr_sk_multilayer_perceptron_net, tprr_sk_multilayer_perceptron_
net, label="AUC= "+ str(area_under_curve_sk_multilayer_perceptron_net))
plt.xlabel("False Positive Rate (FPR)")
plt.ylabel("True Positive Rate (TPR)")
plt.legend(loc="best")
plt.show()
```

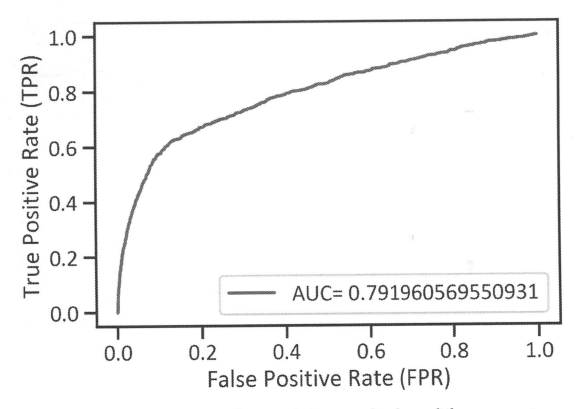

***Figure 7-2.*** *Receiver operating characteristics curve for the multilayer perceptron network executed with the Scikit-Learn framework*

Listing 7-5 arranges the Scikit-Learn multilayer perceptron neural network's precision-recall curve to condense the arrangement of the precision and recall (see Figure 7-3).

**Listing 7-5.** Precision-Recall Curve for the Scikit-Learn Multilayer Perceptron Neural Network

```
p_sk_multilayer_perceptron_net, r_sk_multilayer_perceptron_net, _ =
metrics.precision_recall_curve(y_test, sk_yhat_multilayer_perceptron_net)
weighted_ps_sk_multilayer_perceptron_net = metrics.roc_auc_score(y_test,
sk_yhat_multilayer_perceptron_net)
plt.plot(p_sk_multilayer_perceptron_net, r_sk_multilayer_perceptron_net,
         label="WPR= " +str(weighted_ps_sk_multilayer_perceptron_net))
plt.xlabel("Recall")
plt.ylabel("Precision")
plt.legend(loc="best")
plt.show()
```

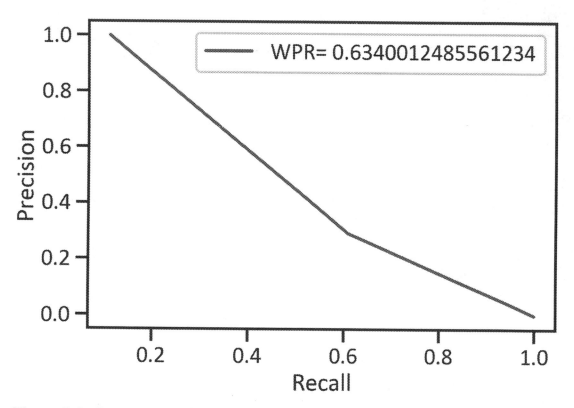

**Figure 7-3.** *Precision-recall curve for the Scikit-Learn multilayer perceptron neural network*

Listing 7-6 arranges the Scikit-Learn multilayer perceptron neural network's learning curve (see Figure 7-4).

***Listing 7-6.*** Arrange a Learning Curve for the Multilayer Perceptron Network Executed with the Scikit-Learn Framework

```
from sklearn.model_selection import learning_curve
train_port_sk_multilayer_perceptron_net, trainscore_sk_multilayer_
perception_net, testscore_sk_multilayer_perceptron_net = learning_
curve(sk_multilayer_perceptron_net, x, y, cv=3, n_jobs=-5, train_sizes=np.
linspace(0.1,1.0,50))
trainscoresk_multilayer_perceptron_net_mean = np.mean(trainscore_sk_
multilayer_perceptron_net, axis=1)
testscoresk_multilayer_perceptron_net_mean = np.mean(testscore_sk_
multilayer_perceptron_net, axis=1)
plt.plot(train_port_sk_multilayer_perceptron_net, trainscoresk_multilayer_
perceptron_net_mean, label="Weighted training accuracy")
plt.plot(train_port_sk_multilayer_perceptron_net, testscoresk_multilayer_
perceptron_net_mean, label="Weighted cv accuracy Score")
plt.xlabel("Training values")
plt.ylabel("Weighted accuracy score")
plt.legend(loc="best")
plt.show()
```

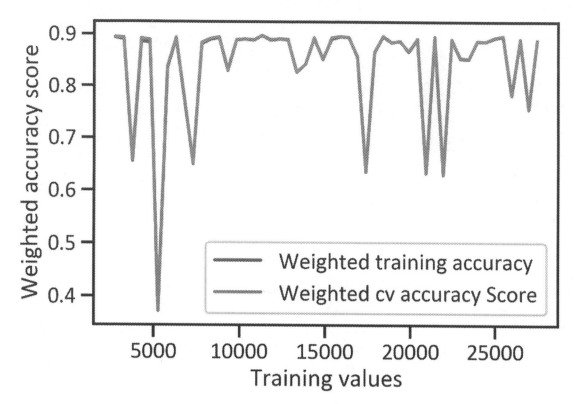

**Figure 7-4.** *Learning curve for the multilayer perceptron network executed with the Scikit-Learn framework*

# Keras in Action

This section executes and assesses a deep belief neural network using the Keras framework. Listing 7-7 preprocesses the features.

*Listing 7-7.* Feature Preprocessing

```
x_train, x_test, y_train, y_test = train_test_split(x, y, test_size=0.2,
random_state=0)
x_train, x_val, y_train, y_val = train_test_split(x_train,y_train,test_
size=0.2,random_state=0)
sk_standard_scaler = StandardScaler()
sk_standard_scaled_x_train = sk_standard_scaler.fit_transform(x_train)
sk_standard_scaled_x_test = sk_standard_scaler.transform(x_test)
```

Listing 7-8 employs the `tensorflow` framework as a backend and installs the Keras framework.

***Listing 7-8.*** Employ the Tensorflow Framework as Backend and Install the Keras Framework

```
import tensorflow as tf
from  tensorflow.keras import Sequential, regularizers
from  tensorflow.keras.layers import Dense
from  tensorflow.keras.wrappers.scikit_learn import KerasClassifier
```

Listing 7-9 structures a multilayer perceptron neural network with two hidden layers that have 17 neurons. The input layer contains a "l1" regularizer with a 0.001 rate, the `relu` activation function, and `binary_crossentropy` loss and accuracy metrics.

***Listing 7-9.*** Structure a Multilayer Perceptron with Keras

```
def keras_multilayer_perceptron_net(optimizer="adam"):
    keras_multilayer_perceptron_net_model = Sequential()
    keras_multilayer_perceptron_net_model.add(Dense(17, input_dim=17,
    activation="sigmoid",kernel_regularizer=regularizers.l1(0.001), bias_
    regularizer=regularizers.l1(0.01)))
    keras_multilayer_perceptron_net_model.add(Dense(17, activation="relu"))
    keras_multilayer_perceptron_net_model.add(Dense(17, activation="relu"))
    keras_multilayer_perceptron_net_model.add(Dense(1, activation="relu"))
    keras_multilayer_perceptron_net_model.compile(loss="binary_
    crossentropy", optimizer=optimizer, metrics=["accuracy"])
    return keras_multilayer_perceptron_net_model
```

Listing 7-10 binds the multilayer perceptron neural network.

***Listing 7-10.*** Bind the Multilayer Perceptron Neural Network

```
keras_multilayer_perceptron_net_model = KerasClassifier(build_fn=keras_
multilayer_perceptron_net)
```

Listing 7-11 executes the Keras multilayer perceptron neural network with 56 epochs and a `batch_size` of 14.

***Listing 7-11.*** Execute the Multilayer Perceptron Neural Network with the Keras Framework

```
keras_multilayer_perceptron_net_model_history = keras_multilayer_
perceptron_net_model.fit(sk_standard_scaled_x_train, y_train, validation_
data=(x_val, y_.val), batch_size=14, epochs=56)
print(keras_multilayer_perceptron_net_model_history)
```

Listing 7-12 arranges a classification report for the multilayer perceptron network executed with the Keras framework (see Table 7-2).

***Listing 7-12.*** Arrange a Classification Report for the Multilayer Perceptron Network Executed with the Keras Framework

```
keras_multilayer_perceptron_net_model.predict(sk_standard_scaled_x_test)
keras_multilayer_perceptron_net_model = pd.DataFrame(metrics.
classification_report(y_test, keras_yhat_multilayer_perceptron_net,
                                    output_dict=True)).transpose()
print(keras_multilayer_perceptron_net_model)
```

***Table 7-2.*** *Classification Report for the Multilayer Perceptron Network Executed with the Keras Framework*

|  | Precision | Recall | F1-score | Support |
|---|---|---|---|---|
| 0 | 0.908443 | 0.991536 | 0.948172 | 7325.000000 |
| 1 | 0.744856 | 0.198248 | 0.313149 | 913.000000 |
| Accuracy | 0.903617 | 0.903617 | 0.903617 | 0.903617 |
| Macro Avg | 0.826649 | 0.594892 | 0.630661 | 8238.000000 |
| Weighted Avg | 0.890313 | 0.903617 | 0.877794 | 8238.000000 |

Listing 7-13 arranges the training and CV loss for a multilayer perceptron network executed with the Keras framework (see Figure 7-5).

**Listing 7-13.** Training and CV Loss for Multilayer Perceptron Network Executed by the Keras Framework

```
plt.plot(keras_multilayer_perceptron_net_model_history.history["loss"],
label="Training loss")
plt.plot(keras_multilayer_perceptron_net_model_history.history["val_loss"],
label="CV loss")
plt.xlabel("Epochs")
plt.ylabel("Loss")
plt.legend(loc="best")
plt.show()
```

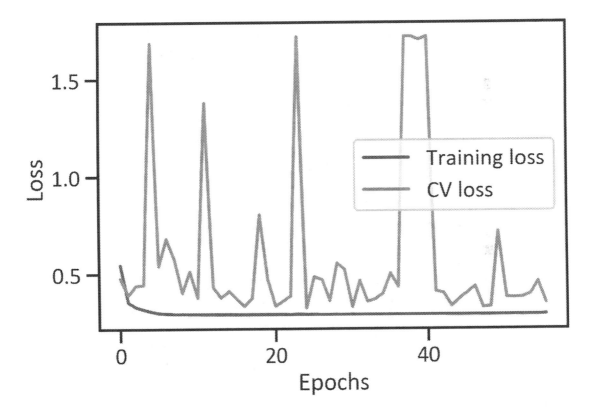

**Figure 7-5.** *Training and CV loss for multilayer perceptron network executed with the Keras framework*

Listing 7-14 arranges the training and CV accuracy for the multilayer perceptron network executed with the Keras framework (see Figure 7-6).

**Listing 7-14.** Arrange the Training and CV Accuracy for Multilayer Perceptron Network Executed with the Keras Framework

```
plt.plot(keras_multilayer_perceptron_net_model_history.history["accuracy"],
label="Training accuracy score")
plt.plot(keras_multilayer_perceptron_net_model_history.history
["val_accuracy"], label="CV accuracy score")
plt.xlabel("Epochs")
plt.ylabel("Accuracy")
plt.legend(loc="best")
plt.show()
```

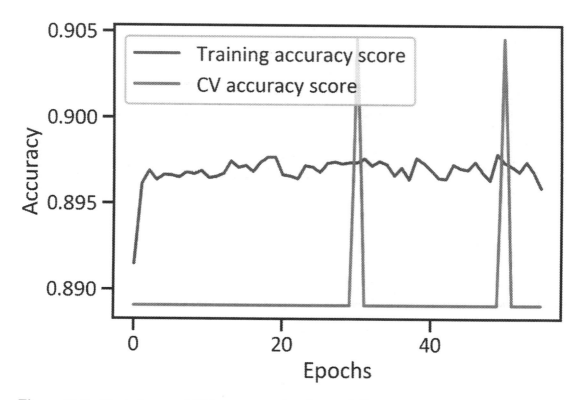

**Figure 7-6.** *Training and CV accuracy for the multilayer perceptron network executed with the Keras framework*

# Deep Belief Networks

Deep belief networks transcend the multilayer perceptron neural network by including hidden layers. The network must model the data repetitively before generating an output value.

# H2O in Action

This section executes a deep belief neural network using the H2O framework.

Listing 7-15 prepares the H2O framework.

***Listing 7-15.*** Prepare the H2O Framework

```
import h2o as initialize_h2o
initialize_h2o.init()
```

Listing 7-16 changes the pandas dataframe to an H2O dataframe.

***Listing 7-16.*** Change the Pandas Dataframe to the H2O Dataframe

```
h2o_data = initialize_h2o.H2OFrame(initial_data)
```

Listing 7-17outlines the features.

***Listing 7-17.*** Outline the Features

```
x_list = list(initial_data.iloc[::, 0:17].columns)
y_list = str(initial_data.columns[-1])
y = y_list
x = h2o_data.col_names
x.remove(y_list)
```

Listing 7-18 randomly divides the data.

***Listing 7-18.*** Randomly Divide the Dataframe

```
h2o_training_data, h2o_validation_data, h2o_test_data = h2o_data.split_
frame(ratios=[.8,.1])
```

Listing 7-19 executes the H2O deep belief neural network.

**Listing 7-19.** Execute the H2O Deep Belief Neural Network

```
from h2o.estimators.deeplearning import H2ODeepLearningEstimator
h2o_deep_belief_net = H2ODeepLearningEstimator(hidden=[5], epochs=56)
h2o_deep_belief_net.train(x= x, y= y, training_frame = h2o_training_data,
validation_frame = h2o_validation_data)h2o_training_data, validation_frame
= h2o_validation_data)
```

# Conclusion

This chapter executed three key machine learning frameworks (Scikit-Learn, Keras, and H2O) to model the data. It produced a binary outcome using the multilayer perceptron and deep belief network. It also applied the binary_crossentropy loss function (to assess the neural networks) and the adam optimizer (to enhance the neural network's performance).

# CHAPTER 8

# Cluster Analysis with Scikit-Learn, PySpark, and H2O

This chapter explains the *k-means* cluster method by implementing a diverse set of Python frameworks (i.e., Scikit-Learn, PySpark, and H2O). To begin, it clarifies how the method apportions values to clusters.

## Exploring the K-Means Method

The k-means method is the most common distance-computing method. It is part of the unsupervised machine learning family. It employs the Euclidean distance objective function to efficiently compute the distance between values, then determines the centers and draws values adjacent to the centers to establish clusters.

Compared to other cluster methods (i.e., spatial cluster methods, birch, and agglomerative), this method requires you to stipulate the number of k before training. This chapter uses the *elbow curve,* which discloses fluctuations in the linear altered vectors to determine the number of $k$. This chapter executes the method to group customers in clusters based on specific characteristics (i.e., age, remuneration, and spending habits).

Listing 8-1 attains the necessary data from a Microsoft CSV file.

***Listing 8-1.*** Attain the Data

```
import pandas as pd
df = pd.read_csv(r"filepath\Mall_Customers.csv")
```

© Tshepo Chris Nokeri 2022
T. C. Nokeri, *Data Science Solutions with Python*, https://doi.org/10.1007/978-1-4842-7762-1_8

Listing 8-2 drops the unnecessary features in the data.

***Listing 8-2.*** Drop Unnecessary Features

```
drop_column_names = df.columns[[0, 1]]
initial_data = df.drop(drop_column_names, axis="columns")
```

Listing 8-3 scales the data with a standard scaler.

***Listing 8-3.*** Scale the Data

```
from sklearn.preprocessing import StandardScaler
sk_standard_scaler = StandardScaler()
sk_standard_scaled_data = sk_standard_scaler.fit_transform(initial_data)
```

Listing 8-4 executes the principal component method using the Scikit-Learn framework.

***Listing 8-4.*** Execute the Principal Component Method with the Scikit-Learn Framework

```
from sklearn.decomposition import PCA
sk_principal_component_method = PCA(n_components=3)
sk_principal_component_method.fit(sk_standard_scaled_data)
sk_principal_component_method_transformed_data = sk_principal_component_
method.transform(sk_standard_scaled_data)
```

Listing 8-5 discloses the number of $k$ using an elbow curve (see Figure 8-1).

***Listing 8-5.*** Disclose the Number of K Using an Elbow Curve

```
from sklearn.cluster import KMeans
import matplotlib.pyplot as plt
%matplotlib inline
elbow_range = range(1, 15)
sk_kmeans_method = [KMeans(n_clusters=x) for x in elbow_range]
sk_kmeans_method_score = [sk_kmeans_method[x].fit(sk_principal_component_
method_transformed_data).score(sk_principal_component_method_transformed_
data) for x in range(len(sk_kmeans_method))]
plt.plot(elbow_range, sk_kmeans_method_score)
```

```
plt.xlabel("Number of k")
plt.ylabel("Eigenvalues")
plt.show()
```

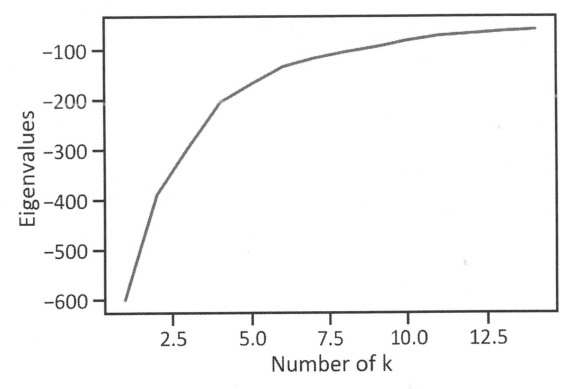

***Figure 8-1.***  *Disclose the number of k using an elbow curve*

Figure 8-1 shows that the curve abruptly twists, which indicates that the k-means method must contain three components.

# Scikit-Learn in Action

This section executes the k-means method using the Scikit-Learn framework; see Listing 8-6.

***Listing 8-6.***  Execute the K-Means Method Using the Scikit-Learn Framework

```
sk_kmeans_method = KMeans(n_clusters=3)
sk_kmeans_method_fitted = sk_kmeans_method.fit(sk_standard_scaled_data)
```

Listing 8-7 computes the Scikit-Learn k-means method's labels.

***Listing 8-7.*** Compute the Scikit-Learn K-Means Method's Labels

```
sk_kmeans_method_labels = pd.DataFrame(sk_kmeans_method_fitted.labels_,
columns = ["Labels"])
```

Listing 8-8 computes the cluster centers and determines their mean and standard deviation (see Table 8-1).

***Listing 8-8.*** Compute Cluster Centers and Their Mean and Standard Deviation

```
sk_kmeans_method = KMeans(n_clusters=3)
sk_kmeans_method_centers = sk_kmeans_method_fitted.cluster_centers_
sk_kmeans_method_centers = pd.DataFrame(sk_kmeans_method_centers, columns =
("1st center","2nd center","3rd center"))
sk_kmeans_method_centers_mean = pd.DataFrame(sk_kmeans_method_centers.
mean())
sk_kmeans_method_centers_std = pd.DataFrame(sk_kmeans_method_centers.std())
sk_kmeans_method_centers_mean_std = pd.concat([sk_kmeans_method_centers_
mean, sk_kmeans_method_centers_std], axis=1)
sk_kmeans_method_centers_mean_std.columns =["Center mean", "Center standard
deviation"]
print(sk_kmeans_method_centers_mean_std)
```

***Table 8-1.*** *Centroid Statistics*

|  | Center Mean | Center Standard Deviation |
| --- | --- | --- |
| First Center | -0.157489 | 0.942836 |
| Second Center | 0.129950 | 0.854012 |
| Third Center | 0.222984 | 0.891719 |

Listing 8-9 determines the scaled data and labels that the Scikit-Learn k-means method computed (see Figure 8-2).

***Listing 8-9.*** Determine the PySpark K-Means Method 's Labels

```
plt.scatter(sk_principal_component_method_transformed_data[:,0],
sk_principal_component_method_transformed_data[:, 1],
          c=sk_kmeans_method_fitted.labels_,cmap="coolwarm", s=120)
plt.xlabel("y")
plt.show()
```

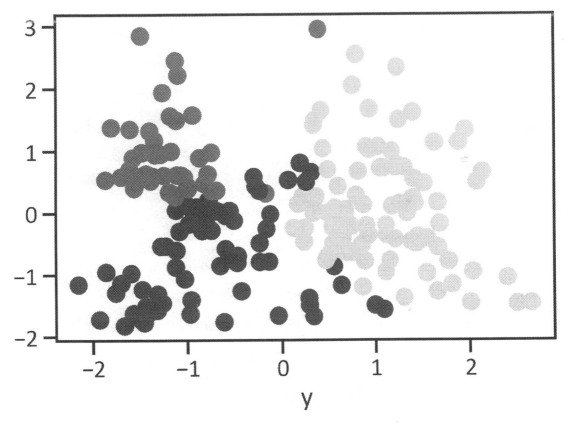

***Figure 8-2.*** *Scikit-Learn k-means method labels*

# PySpark in Action

This section executes and assesses the k-means method using the PySpark framework.
Listing 8-10 prepares the PySpark framework using the findspark framework.

***Listing 8-10.*** Prepare the PySpark Framework

```
import findspark as initiate_pyspark
initiate_pyspark.init("filepath\spark-3.0.0-bin-hadoop2.7")
```

Listing 8-11 stipulates the PySpark app using the SparkConf() method.

***Listing 8-11.*** Stipulate the PySpark App

```
from pyspark import SparkConf
pyspark_configuration = SparkConf().setAppName("pyspark_kmeans_method").
setMaster("local")
```

Listing 8-12 prepares the PySpark session using the SparkSession() method.

***Listing 8-12.*** Prepare the Spark Session

```
from pyspark.sql import SparkSession
pyspark_session = SparkSession(pyspark_context)
from pyspark.sql import SparkSession
```

Listing 8-13 changes the pandas dataframe created earlier in this chapter to a PySpark dataframe using the createDataFrame() method.

***Listing 8-13.*** Change the Pandas Dataframe to a PySpark Dataframe

```
pyspark_initial_data = pyspark_session.createDataFrame(initial_data)
```

Listing 8-14 creates a list for the independent features and a string for the dependent feature. It then converts the data using the VectorAssembler() method for modeling with the PySpark framework.

***Listing 8-14.*** Transform the Data

```
x_list = list(initial_data.iloc[::, 0:3].columns)
from pyspark.ml.feature import VectorAssembler
pyspark_data_columns = x_list
pyspark_vector_assembler = VectorAssembler(inputCols=pyspark_data_columns,
outputCol="features")
pyspark_data = pyspark_vector_assembler.transform(pyspark_initial_data)
```

Listing 8-15 executes the k-means method using the PySpark framework.

***Listing 8-15.*** Execute the K-Means Method Using the PySpark Framework

```
from pyspark.ml.clustering import KMeans
pyspark_kmeans_method = KMeans().setK(3).setSeed(1)
pyspark_kmeans_method_fitted = pyspark_kmeans_method.fit(pyspark_data)
```

Listing 8-16 computes the k-means method's labels and passes them to a pandas dataframe.

***Listing 8-16.*** Compute the PySpark K-Means Method 's Labels

```
pyspark_yhat = pyspark_kmeans_method_fitted.transform(pyspark_data)
pyspark_yhat_pandas_df = pyspark_yhat.toPandas()
```

Listing 8-17 shows the scaled data and labels that the PySpark k-means method computed (see Figure 8-3).

***Listing 8-17.*** Disclose the PySpark K-Means Method's Labels

```
from mpl_toolkits.mplot3d import Axes3D
figure = plt.figure(figsize=(12, 12))
ax = Axes3D(figure)
ax.scatter(pyspark_yhat_pandas_df.iloc[::, 0],
           pyspark_yhat_pandas_df.iloc[::, 1],
           pyspark_yhat_pandas_df.iloc[::, 2],
           c = pyspark_yhat_pandas_df.iloc[::, 4],
           cmap = "coolwarm",
           s = 120)
ax.set_xlabel(initial_data.columns[0])
ax.set_ylabel(initial_data.columns[1])
ax.set_zlabel(initial_data.columns[2])
figure.show()
```

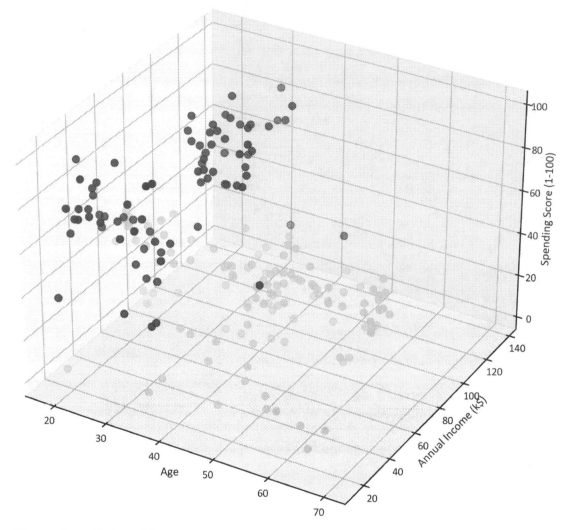

***Figure 8-3.*** *PySpark k-means method labels*

Listing 8-18 computes the PySpark k-means method's centers.

***Listing 8-18.*** Compute the PySpark K-Means Method's Centers

```
pyspark_kmeans_method_centers = pyspark_kmeans_method_fitted.
clusterCenters()
print("K-Means method cluster centers")
for pyspark_cluster_center in pyspark_kmeans_method_centers:
    print(pyspark_cluster_center)
K-Means method cluster centers
```

```
[26.5952381   33.14285714 65.66666667]
[45.30769231 60.52991453 34.47008547]
[32.97560976 88.73170732 79.24390244]
```

Listing 8-19 assesses the PySpark k-means method using the Silhouette method.

**Listing 8-19.** Assess the PySpark K-Means Method Using the Silhouette Method

```
from pyspark.ml.evaluation import ClusteringEvaluator
pyspark_kmeans_method_assessment = ClusteringEvaluator()
pyspark_kmeans_method_silhoutte = pyspark_kmeans_method_assessment.
evaluate(pyspark_yhat)
print("Silhouette score = " + str(pyspark_kmeans_method_silhoutte))

Silhouette score = 0.45097978411312134
```

# H2O in Action

This section executes and assesses the k-means method using the H2O framework.

Listing 8-20 prepares the H2O framework.

**Listing 8-20.** Prepare the H2O Framework

```
import h2o as initialize_h2o
initialize_h2o.init()
```

Listing 8-21 changes the pandas dataframe to an H2O dataframe .

**Listing 8-21.** Change the Pandas Dataframe to an H2O Dataframe

```
h2o_data = initialize_h2o.H2OFrame(initial_data)
```

Listing 8-22 outlines the independent and dependent features.

**Listing 8-22.** Outline the Features

```
y = y_list
x = h2o_data.col_names
x.remove(y_list)
```

Listing 8-23 reserves data for training and validating the H2O k-means method.

**Listing 8-23.** Randomly Divide the Dataframe

```
h2o_training_data, h2o_validation_data = h2o_data.split_frame(ratios=[.8])
```

Listing 8-24 executes the H2O k-means method.

**Listing 8-24.** Execute the K-Means Method Using the H2O Framework

```
from h2o.estimators import H2OKMeansEstimator
h2o_kmeans_method = H2OKMeansEstimator(k=3)
h2o_kmeans_method.train(x = x_list, training_frame = h2o_training_data,
validation_frame = h2o_validation_data)
```

Listing 8-25 computes the H2O k-means method's labels.

**Listing 8-25.** Compute the H2O K-Means Method Labels

```
h2o_yhat = h2o_kmeans_method.predict(h2o_validation_data)
```

Listing 8-26 assesses the H2O k-means method (see Table 8-2).

**Listing 8-26.** Assess the K-Means Method Executed with the H2O Framework

```
from h2o.estimators import H2OKMeansEstimator
h2o_kmeans_method_assessment = h2o_kmeans_method.model_performance()
h2o_kmeans_method_assessment

ModelMetricsClustering: kmeans
** Reported on train data. **
MSE: NaN
RMSE: NaN
Total Within Cluster Sum of Square Error: 238.6306804611187
Total Sum of Square Error to Grand Mean: 479.99999503427534
Between Cluster Sum of Square Error: 241.36931457315666
```

***Table 8-2.*** *PySpark Centroid Statistics*

|   | Centroid | Size | Within_cluster_sum_of_squares |
|---|----------|------|-------------------------------|
| 0 | 1.0 | 78.0 | 134.983658 |
| 1 | 2.0 | 51.0 | 69.748489 |
| 2 | 3.0 | 32.0 | 33.898533 |

# Conclusion

This chapter executed three key machine learning frameworks (Scikit-Learn, PySpark, and H2O) in order to group data into three clusters. You saw how to identify the number of $k$ using the elbow curve. To assess the method, you employed the Silhouette method and looked at the sum of squared errors.

# CHAPTER 9

# Principal Component Analysis with Scikit-Learn, PySpark, and H2O

This chapter executes a simple dimension reducer (a principal component method) by implementing a diverse set of Python frameworks (Scikit-Learn, PySpark, and H2O). To begin, it clarifies how the method computes components.

## Exploring the Principal Component Method

The principal component method is a simple dimension reducer. It carries out linear transformations on the entire data set to attain vectors (identified as *eigenvalues*), then identifies incremental changes in the data. It enables you to detect changes in features in the data. This method helps determine the ratio of each feature so you can decide which features to preserve for modeling. It is suitable when your data contains countless features, because it helps you identify features that are imperative to the model. It is important to note that, although I call it a "method" in this chapter, that is not strictly the case. Compared to other methods, it does not consider any plausible hypotheses; rather, you execute it prior to executing linear, nonlinear, and cluster methods, among others.

This chapter uses the data from the previous chapter, which covered the k-means method.

Listing 9-1 attains the necessary data from a Microsoft CSV file.

© Tshepo Chris Nokeri 2022
T. C. Nokeri, *Data Science Solutions with Python*, https://doi.org/10.1007/978-1-4842-7762-1_9

***Listing 9-1.*** Attain the Data

```
import pandas as pd
initial_data = df.drop(drop_column_names, axis="columns")
```

Listing 9-2 drops any unnecessary features in the data.

***Listing 9-2.*** Drop Unnecessary Features

```
drop_column_names = df.columns[[0, 1]]
initial_data = df.drop(drop_column_names, axis="columns")
```

# Scikit-Learn in Action

Listing 9-3 scales the whole data set with a standard scaler.

***Listing 9-3.*** Scale the Whole Data Set

```
from sklearn.preprocessing import StandardScaler
sk_standard_scaler = StandardScaler()
sk_standard_scaled_data = sk_standard_scaler.fit_transform(initial_data))
```

Listing 9-4 executes the principal component method with the Scikit-Learn framework. It then discloses the variance that each component clarifies (see Figure 9-1).

***Listing 9-4.*** Arrange the Explained Variance from the Scikit-Learn Principal Components Method

```
import matplotlib.pyplot as plt
%matplotlib inline
from sklearn.decomposition import PCA
sk_principal_component_method = PCA(n_components=3)
sk_principal_component_method.fit_transform(sk_standard_scaled_data)
sk_principal_component_method_variance = sk_principal_component_method.
explained_variance_
plt.bar(range(3), sk_principal_component_method_variance, label="Variance")
plt.legend()
```

```
plt.ylabel("Variance ratio")
plt.xlabel("Components")
plt.show()
```

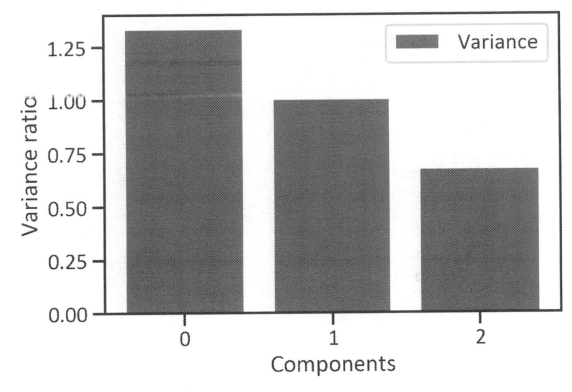

***Figure 9-1.*** *Variance of the Scikit-Learn principal components method*

Figure 9-1 shows all that components disclose over 0.5 of variance. Therefore, all the principal component methods will have three components.

Listing 9-5 displays the variance from the Scikit-Learn principal component method.

***Listing 9-5.*** Print the Variance that the Scikit-Learn Principal Component Method Discloses

```
print("SciKit-Learn PCA explained variance ratio", sk_principal_component_
method_variance)

SciKit-Learn PCA explained variance ratio [1.33465831 1.00427272
0.67614435]
```

Listing 9-6 executes the principal component method with the Scikit-Learn framework and shows fewer dimensions (see Figure 9-2).

**Listing 9-6.** Execute the Principal Component Method with the Scikit-Learn Framework

```
sk_principal_component_method_ = PCA(n_components=3)
sk_principal_component_method_.fit(sk_standard_scaled_data)
sk_principal_component_method_reduced_data = sk_principal_component_
method_.transform(sk_standard_scaled_data)
plt.scatter(sk_principal_component_method_reduced_data[:,0], sk_
principal_component_method_reduced_data[:,2], c=initial_data.iloc[::, -1],
cmap="coolwarm")
plt.xlabel("y")
plt.show()
```

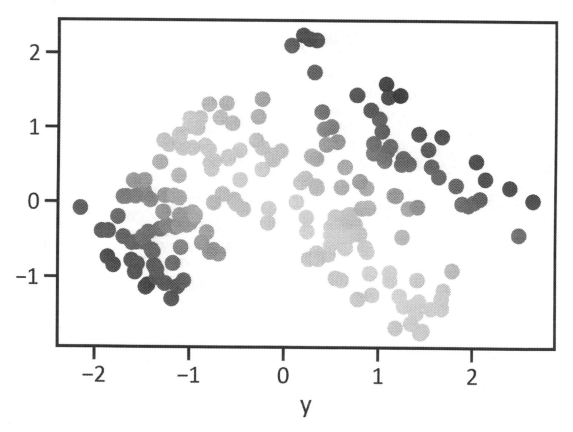

**Figure 9-2.** *Dimensions from the Scikit-Learn principal component method*

Figure 9-2 shows the dimensions found by the Scikit-Learn principal component method.

# PySpark in Action

This section executes the principal component method using the PySpark framework.

Listing 9-7 prepares the PySpark framework using the findspark framework.

**Listing 9-7.** Prepare the PySpark Framework

```
import findspark as initiate_pyspark
initiate_pyspark.init("filepath\spark-3.0.0-bin-hadoop2.7")
```

Listing 9-8 stipulates the PySpark app using the SparkConf() method.

**Listing 9-8.** Stipulate the PySpark App

```
from pyspark import SparkConf
pyspark_configuration = SparkConf().setAppName("pyspark_principal_
components_analysis").setMaster("local")
```

Listing 9-9 prepares the PySpark session using the SparkSession() method.

**Listing 9-9.** Prepare the Spark Session

```
from pyspark.sql import SparkSession
pyspark_session = SparkSession(pyspark_context)
from pyspark.sql import SparkSession
```

Listing 9-10 changes the pandas dataframe created earlier in this chapter to a PySpark dataframe using the createDataFrame() method.

**Listing 9-10.** Change the Pandas Dataframe to a PySpark Dataframe

```
pyspark_initial_data = pyspark_session.createDataFrame(initial_data)
```

Listing 9-11 creates a list for the independent features and a string for the dependent feature. It then converts the data using the VectorAssembler() method for modeling with the PySpark framework.

***Listing 9-11.*** Transform the Data

```
x_list = list(initial_data.iloc[::, 0:3].columns)
from pyspark.ml.feature import VectorAssembler
pyspark_data_columns = x_list
pyspark_vector_assembler = VectorAssembler(inputCols=pyspark_data_columns,
outputCol="variables")
pyspark_data = pyspark_vector_assembler.transform(pyspark_initial_data)
```

Listing 9-12 scales the whole data set with the Lifeline PySpark framework.

***Listing 9-12.*** Scale the Whole Data Set with the Lifeline PySpark Framework

```
from pyspark.ml.feature import StandardScaler
pyspark_standard_scaler = StandardScaler(inputCol = pyspark_data.
columns[-1], outputCol = "pyspark_scaled_features", withMean = True,
withStd = True).fit(pyspark_data)
pyspark_data_scaled_data = pyspark_standard_scaler.transform(pyspark_data)
```

Listing 9-13 executes the PySpark principal component method.

***Listing 9-13.*** Execute the PySpark Principal Component Method

```
from pyspark.ml.feature import PCA
number_of_k = 3
pyspark_principal_components_method = PCA(k = number_of_k, inputCol =
pyspark_data_scaled_data.columns[-1], outputCol = "pca_reduced_features").
fit(pyspark_data_scaled_data)
pyspark_principal_components_method_data = pyspark_principal_components_
method.transform(pyspark_data_scaled_data)
```

Listing 9-14 discloses the variance that each component clarifies (see Figure 9-3).

***Listing 9-14.*** Arrange the PySpark Principal Components Method's Explained Variance

```
pyspark_principal_component_method_variance = pyspark_principal_components_
method.explainedVariance.toArray()
plt.bar(range(3), pyspark_principal_component_method_variance,
label="Variance")
```

```
plt.legend()
plt.ylabel("Variance ratio")
plt.xlabel("Components")
plt.show()
```

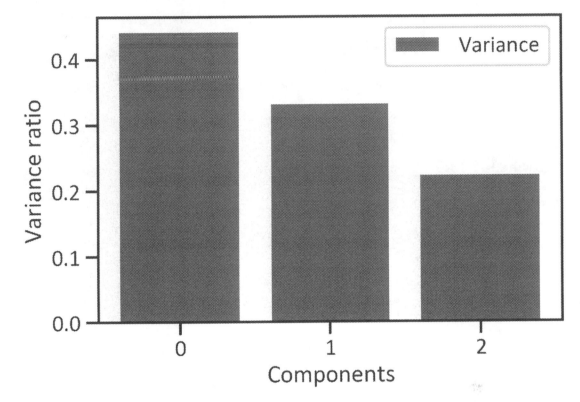

***Figure 9-3.*** *Explained variance from the PySpark principal components method*

Listing 9-15 prints the variance that the PySpark principal component method reveals.

***Listing 9-15.*** Print the Variance that the PySpark Principal Component Method Reveals

```
print('Explained variance ratio', pyspark_principal_components_method.
explainedVariance.toArray())
PySpark PCA variance ratio [0.44266167 0.33308378 0.22425454]
```

The results show that the first component discloses 0.44 of the related changes in features in the data, the second component discloses 0.33, and the third component discloses 0.22.

Listing 9-16 reduces the data using the PySpark principal component method.

***Listing 9-16.*** Reduce the Data with the PySpark Principal Component Method

```
pyspark_principal_components_method_reduced_data = pyspark_principal_
components_method_data.rdd.map(lambda row: row.pyspark_scaled_features).
collect()
pyspark_principal_components_method_reduced_data = np.array(pyspark_
principal_components_method_reduced_data)
```

Listing 9-17 discloses the data that the PySpark principal component method reduced (see Figure 9-4).

***Listing 9-17.*** Disclose the Data that the PySpark Principal Component Method Reduced

```
plt.scatter(pyspark_principal_components_method_reduced_data[:,0], pyspark_
principal_components_method_reduced_data[:,1], c=initial_data.iloc[::, -1],
cmap = "coolwarm")
plt.xlabel("y")
plt.show()
```

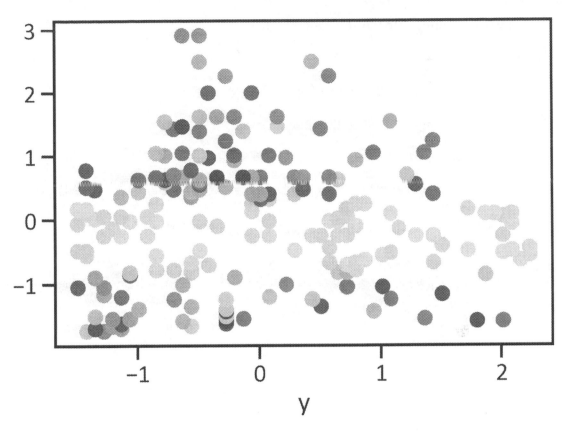

***Figure 9-4.*** *Dimensions from the PySpark principal component method*

Figure 9-4 shows the dimensions found by the PySpark principal component method.

## H2O in Action

This section executes and assesses the principal component method using the H2O framework.

Listing 9-18 prepares the H2O framework.

***Listing 9-18.*** Prepare the H2O Framework

```
import h2o as initialize_h2o
initialize_h2o.init()
```

Listing 9-19 changes the pandas dataframe to an H2O dataframe.

***Listing 9-19.*** Change the Pandas Dataframe to an H2O Dataframe

```
h2o_data = initialize_h2o.H2OFrame(initial_data)
```

Listing 9-20 randomly divides the dataframe.

***Listing 9-20.*** Randomly Divide the Dataframe

```
h2o_training_data, h2o_validation_data = h2o_data.split_frame(ratios=[.8])
```

Listing 9-21 executes the H2O principal component method.

***Listing 9-21.*** Execute the H2O Principal Component Method

```
from h2o.estimators import H2OPrincipalComponentAnalysisEstimator
h2o_principal_components_method = H2OPrincipalComponentAnalysisEstimator
(k = 3, transform = "standardize")
h2o_principal_components_method.train(training_frame = h2o_training_data)
```

Listing 9-22 reduces the data with the H2O principal component method.

***Listing 9-22.*** Reduce the Data with the H2O Principal Component Method

```
h2o_yhat = h2o_principal_components_method.predict(h2o_validation_data)
```

# Conclusion

This chapter executed three key machine learning frameworks (Scikit-Learn, PySpark, and H2O) in order to condense data into a few dimensions by employing the principal component method. As mentioned at the beginning of this chapter, the chapter did not execute the method to test plausible hypotheses, thus there are no key metrics for objectively assessing the method.

# CHAPTER 10

# Automating the Machine Learning Process with H2O

This is a short yet insightful chapter that concludes the book by explaining a straightforward approach to automating machine learning processes with the help of a widespread machine learning framework, known as H2O.

## Exploring Automated Machine Learning

In a world saturated by innumerable machine learning frameworks, you might think that organizations would easily adopt machine learning. Unfortunately, that is not the case. There is a massive shortage of skills because of the considerable complexity of machine algorithms.

To combat these challenges, H2O introduced AutoML, which attempts to automate the machine learning process by simultaneously training and handpicking methods with extraordinary performance.

Although the H2O AutoML function is beneficial, it does not include methods like cluster analysis, survival analysis, and time series analysis, among others.

This chapter uses the data from Chapter 3. It also distinguishes the performance of the leading AutoML method against the ones executed in Chapter 3.

111

© Tshepo Chris Nokeri 2022
T. C. Nokeri, *Data Science Solutions with Python*, https://doi.org/10.1007/978-1-4842-7762-1_10

# Preprocessing Features

This chapter manipulates the data attained in Chapter 3, so it does not describe the preprocessing tasks in detail. Listing 10-1 executes all the preprocessing tasks.

***Listing 10-1.*** Preprocess Features

```
import pandas as pd
df = pd.read_csv(r"filepath\WA_Fn-UseC_-Marketing_Customer_Value_Analysis.csv")
drop_column_names = df.columns[[0, 6]]
initial_data = df.drop(drop_column_names, axis="columns")
initial_data.iloc[::, 0] = pd.get_dummies(initial_data.iloc[::, 0])
initial_data.iloc[::, 2] = pd.get_dummies(initial_data.iloc[::, 2])
initial_data.iloc[::, 3] = pd.get_dummies(initial_data.iloc[::, 3])
initial_data.iloc[::, 4] = pd.get_dummies(initial_data.iloc[::, 4])
initial_data.iloc[::, 5] = pd.get_dummies(initial_data.iloc[::, 5])
initial_data.iloc[::, 6] = pd.get_dummies(initial_data.iloc[::, 6])
initial_data.iloc[::, 7] = pd.get_dummies(initial_data.iloc[::, 7])
initial_data.iloc[::, 8] = pd.get_dummies(initial_data.iloc[::, 8])
initial_data.iloc[::, 9] = pd.get_dummies(initial_data.iloc[::, 9])
initial_data.iloc[::, 15] = pd.get_dummies(initial_data.iloc[::, 15])
initial_data.iloc[::, 16] = pd.get_dummies(initial_data.iloc[::, 16])
initial_data.iloc[::, 17] = pd.get_dummies(initial_data.iloc[::, 17])
initial_data.iloc[::, 18] = pd.get_dummies(initial_data.iloc[::, 18])
initial_data.iloc[::, 20] = pd.get_dummies(initial_data.iloc[::, 20])
initial_data.iloc[::, 21] = pd.get_dummies(initial_data.iloc[::, 21])
```

# H2O AutoML in Action

This section uses the H2OAutoML() method to train multiple methods and rank their performance. This section executes and assesses H2OAutoML.

Listing 10-2 prepares the H2O framework.

***Listing 10-2.*** Prepare the H2O Framework

```
import h2o as initialize_h2o
initialize_h2o.init()
```

Listing 10-3 changes the pandas dataframe to the H2O dataframe.

***Listing 10-3.*** Change the Pandas Dataframe to H2O Dataframe

```
h2o_data = initialize_h2o.H2OFrame(initial_data)
```

Listing 10-4 outlines the independent and dependent features.

***Listing 10-4.*** Outline the Features

```
int_x = initial_data.iloc[::,0:19]
fin_x = initial_data.iloc[::,19:21]
x_combined = pd.concat([int_x, fin_x], axis=1)
x_list = list(x_combined.columns)
y_list = initial_data.columns[19]
y = y_list
x = h2o_data.col_names
x.remove(y_list)
```

Listing 10-5 randomly divides the data.

***Listing 10-5.*** Randomly Divide the Dataframe

```
h2o_training_data, h2o_validation_data, h2o_test_data = h2o_data.split_
frame(ratios=[.8,.1])
```
The H2OAutoML() method enables us to stipulate the considerable time needed to train each method. Specify max_runtime_secs to do so.

Listing 10-6 executes H2O's auto machine learning method.

***Listing 10-6.*** Execute H2O's Auto Machine Learning

```
from h2o.automl import H2OAutoML
h2o_automatic_ml = H2OAutoML(max_runtime_secs = 240)
h2o_automatic_ml.train(x= x, y= y, training_frame = h2o_training_data,
validation_frame = h2o_validation_data)
```

Listing 10-7 ranks H2O's methods in ascending order (see Table 10-1).

**Listing 10-7.** Rank H2O's Methods

```
h2o_method_ranking = h2o_automatic_ml.leaderboard
print(h2o_method_ranking)
```

**Table 10-1.**  *H2O Method Ranking*

| Model_ID | Mean_residual_ deviance | Rmse | Mse | Mae | Rmsle |
|---|---|---|---|---|---|
| StackedEnsemble_AllModels_ AutoML_20210824_115721 | 18371.3 | 135.541 | 18371.3 | 86.6257 | 0.482043 |
| StackedEnsemble_ BestOfFamily_AutoML_ 20210824_115721 | 18566.8 | 136.26 | 18566.8 | 87.7959 | 0.477828 |
| GBM_1_AutoML_20210824_ 115721 | 19392.8 | 139.258 | 19392.8 | 90.0652 | 0.490441 |
| GBM_3_AutoML_20210824_ 115721 | 19397.8 | 139.276 | 19397.8 | 90.3467 | 0.490855 |
| GBM_2_AutoML_20210824_ 115721 | 19467.3 | 139.525 | 19467.3 | 89.9908 | 0.491887 |
| GBM_grid__1_AutoML_ 20210824_115721_model_4 | 19607.9 | 140.028 | 19607.9 | 91.3311 | 0.495784 |
| GBM_4_AutoML_20210824_ 115721 | 19633.4 | 140.119 | 19633.4 | 89.9553 | 0.499693 |
| XRT_1_AutoML_20210824_ 115721 | 19717.6 | 140.42 | 19717.6 | 89.7474 | 0.485144 |
| GBM_grid__1_AutoML_ 20210824_115721_model_2 | 19805.8 | 140.733 | 19805.8 | 91.7092 | 0.494005 |
| GBM_grid__1_AutoML_ 20210824_115721_model_9 | 19858.6 | 140.921 | 19858.6 | 90.0839 | 0.521174 |

Table 10-1 suggests that StackedEnsemble_AllModels_AutoML_20210824_115721 is the highest-ranking method with the smallest root mean residual deviance (at 18371.3). The root-mean-square error is 135.541.

In contrast, the lowest ranking method is GBM_grid__1_AutoML_20210824_11572 1_model_9, with a mean residual deviance at 19858.6 and a root-mean-square error at 140.921. Listing 10-8 assesses the highest ranking method (StackedEnsemble_ AllModels_AutoML_20210824_115721).

***Listing 10-8.*** Execute H2O Auto Machine Learning

```
highest_ranking_method = h2o_automatic_ml.leader
print(highest_ranking_method)

ModelMetricsRegressionGLM: stackedensemble
** Reported on train data. **

MSE: 4900.131929292816
RMSE: 70.00094234574857
MAE: 45.791336342731775
RMSLE: 0.3107457400431741
R^2: 0.9421217261180531
Mean Residual Deviance: 4900.131929292816
Null degrees of freedom: 7288
Residual degrees of freedom: 7276
Null deviance: 617106545.1167994
Residual deviance: 35717061.632615335
AIC: 82648.0458245655

ModelMetricsRegressionGLM: stackedensemble
** Reported on validation data. **

MSE: 17967.946605287365
RMSE: 134.04456947331872
MAE: 83.66742154264892
RMSLE: 0.43883929226519075
R^2: 0.7718029372846945
Mean Residual Deviance: 17967.946605287365
Null degrees of freedom: 930
```

```
Residual degrees of freedom: 918
Null deviance: 73564698.35450743
Residual deviance: 16728158.289522538
AIC: 11790.460469478105

ModelMetricsRegressionGLM: stackedensemble
** Reported on cross-validation data. **

MSE: 18371.339394332357
RMSE: 135.5409140973026
MAE: 86.62574587341507
RMSLE: 0.4820432203973515
R^2: 0.7830055540572105
Mean Residual Deviance: 18371.339394332357
Null degrees of freedom: 7288
Residual degrees of freedom: 7275
Null deviance: 617269595.4214188
Residual deviance: 133908692.84528854
AIC: 92282.67565857261
```

The results suggest that the StackedEnsemble_AllModels_AutoML_20210824_115721 method explained 77% of the related changes in the test data and 78% of the related changes in the validation.

# Conclusion

The highest-ranking H2O AutoML method (StackedEnsemble_AllModels_ AutoML_20210824_115721) pales in comparison to the ordinary least-square methods executed in the chapter. The H20 method disclosed 77% of the related changes in the test data. In contrast, all methods executed in Chapter 3 (with the Scikit-Learn and PySpark frameworks) explained 100% of the related changes in the test data. Unfortunately, the StackedEnsemble_AllModels_AutoML_20210824_115721 method does not have a summary function. This concludes the book.

# Index

## A

Accelerated failure time method, 34
Apache Spark, 7, 10
AutoML method, 111
    H2O, 112, 113, 115, 116
    preprocessing tasks, 112

## B

Big data, 7
    business, 8
    customer relationships, 8
    decision making, 9
    ETL, 10
    features, 8
    product development, 9
    warehousing, 9

## C

Categories, 2
Cluster centers, 92
Cluster methods, 3, 4
Cox Proportional Hazards method, 29, 30

## D

Data-driven organizations, 9
Decision tree, 59, 60, 62
Deep belief networks, 87

## E

Elbow curve, 89, 91
Ensemble methods, 3

## F

findspark framework, 69

## G

Gaussian distribution, 1
Gradient boosting methods, 66
    H2O framework, 71, 72
    PySpark, 69, 71
    XGBoost, 66–68
GraphX, 12

## H, I, J

H2OAutoML, 112
H2O framework, 22, 23, 25–27, 52, 71
H2O logistic regression
    method, 54, 55, 57
Hadoop Distributed File System (HDF), 11
Hadoop File System (HDFS), 10

Deep learning (DL), 4, 7, 75
Dimension reducers, 4
DL frameworks, Keras, 14

117

Printed in the United States
by Baker & Taylor Publisher Services